THE QUANTUM MATRIX

THE QUANTUM MATRIX

Henry Bar's Perilous Struggle for
Quantum Coherence

GERSHON KURIZKI AND GOREN GORDON

ILLUSTRATED BY ETZION GOEL

OXFORD
UNIVERSITY PRESS

OXFORD

UNIVERSITY PRESS

Great Clarendon Street, Oxford, OX2 6DP,
United Kingdom

Oxford University Press is a department of the University of Oxford.
It furthers the University's objective of excellence in research, scholarship,
and education by publishing worldwide. Oxford is a registered trade mark of
Oxford University Press in the UK and in certain other countries

First Edition published in 2020

Impression: 1

Published in the United States of America by Oxford University Press
198 Madison Avenue, New York, NY 10016, United States of America

British Library Cataloguing in Publication Data
Data available

Library of Congress Control Number: 2020933395

ISBN 978–0–19–878746–4

Printed and bound by
CPI Group (UK) Ltd, Croydon, CR0 4YY

THE QUANTUM MATRIX

HENRY BAR'S PERILOUS STRUGGLE
FOR QUANTUM COHERENCE

GERSHON **KURIZKI** AND GOREN **GORDON**

ILLUSTRATED BY ETZION **GOEL**

OXFORD

We dedicate this book to Zipi, Michal and Yaarit, whose indulgence and support have made it possible. We are grateful to David Petrosyan for his invaluable comments and critique.

PREFACE

"Meet Henry Bar, a physicist and . . . quantum superhero". The title *The Quantum Matrix* refers to a central concept in quantum physics, but also (allegorically) to our enigmatic world. In this book Henry Bar, physicist and the first quantum superhero, guides the reader through the amazing quantum world. Henry's hair-raising adventures in his perilous struggle for quantum coherence are graphically depicted by comics and thoroughly explained to the lay reader. Behind each adventure lies a key concept in quantum physics. These concepts range from the basic quantum coherence and entanglement through tunneling and the recently discovered quantum decoherence control, to the principles of the emerging technologies of quantum communication and computing. The explanations of the concepts are popular, but nonetheless rigorous and detailed, and are followed by an account of the broader context of these concepts, their historic perspective, up-to-date status and forthcoming developments. Finally, thought-provoking philosophical and cultural implications of these concepts are discussed. The mathematical appendices of all chapters cover, in a straightforward manner, the core aspects of quantum physics at the level of a university introductory course.

The Quantum Matrix is composed of sections that are ordered as the elements of a matrix, which is a basic tool in quantum mechanics. Each chapter of the book is a "row" in this matrix (see Matrix of Contents) that describes a key theme of the quantum world. In turn, each chapter contains a comic-strip episode followed by four sections. The comic-strip episode is intended for readers of all ages and backgrounds. It tells of the thrilling adventures of Henry Bar, a quantum physicist who has mastered the use of quantum effects to perfection, so as to become a quantum superhero. He perilously struggles to defend his friends, Alice and Bob, against their foe Eve, a powerful and sometimes ruthless opponent who employs classical effects for her mischief or pranks. As this tale of suspense unfolds through consecutive chapters, the reader is confronted with more and more complex issues of quantum mechanics. The surprising ending of the story puts these issues in common perspective.

The first factual section after the comic strip in each chapter explains the physics underlying Henry Bar's feats at the level of popular science. The second factual section of each chapter concisely describes the development of the concepts that feature in the corresponding episode and relates them to real-world experiments and applications. The third section of each chapter contemplates a philosophical issue related to the effect discussed. Each chapter also includes a "quantum verse" that muses over the issues raised.

The appendix to each chapter introduces the reader to the basic mathematical description of the quantum effect in question in a simple and straightforward manner. The reader does not need prior knowledge—only the willingness to delve into the intricacies of Henry's adventures that are described by the equations of quantum physics that govern our world. Each appendix allows a deeper insight into the narrative and the underlying physics. If the appendices of all chapters are read consecutively, they form an elementary introduction to quantum physics for pedestrians, starting from the most elementary notions and culminating in advanced topics.

The book can be read in two distinct ways, either along its columns (sections) or along its rows (chapters). Reading it along the columns caters to different readerships.

- A reader who has not yet encountered the wondrous world of quantum physics may choose to first become acquainted with it through the adventures of Henry Bar, by reading the comic strip of each chapter and postponing the reading of the other sections until later.
- Aficionados of popular science and science history will find, in the first two factual sections of each chapter, a succinct, informative, non-mathematical description of quantum physics from both historical and contemporary perspectives.
- The third section is recommended to everyone—especially to those that are fond of relating the "big" questions of science to the broad cultural heritage of mankind and obtaining unorthodox perspectives of quantum physics.
- The appendix is intended for those who dare take a first plunge into the basic mathematics of quantum mechanics.

The other, more conventional, way is to read the book chapter by chapter, starting from simple notions and progressing towards more advanced ideas. For those who already know some quantum physics and have the basic mathematical background, this is the recommended way to read.

The rows of *The Quantum Matrix*—its chapters—form Parts I–III of the book. Part I, consisting of Chapter 1–6, is concerned with the basic concepts of quantum mechanics (QM).

- Chapter 1 surveys a wide range of quantum phenomena, trying to impress upon the reader the importance of quantum physics in understanding the world, the current technological advances and "the way of the future", which, in our view, will be heavily dominated by quantum principles.

- Chapter 2–3 introduce the basic notions of quantum superposition and interference.
- Chapter 4 describes the unique role of measurements in quantum mechanics.
- Chapter 5–6 explaore the uncertainty principle, first introducing the position–momentum uncertainty and then the time–energy uncertainty.

In Part II, consisting of Chapter 7–12, the emphasis is placed on the so-called "open-system effects"; i.e., effects related to the interaction of quantum systems with their environment.

- Chapter 7 introduces the key concept of quantum entanglement, the basic element of multi-partite systems.
- Chapter 8 discusses the relation between entanglement and decoherence, along with the notions of decoherence as "which-path" information and its "quantum eraser" control.
- Chapter 9 describes environment effects that hamper superposition and entanglement; namely, dephasing (decoherence) and decay (relaxation).
- Chapter 10 introduces dynamical control of the evolution by means of the Zeno and anti-Zeno effects.
- Chapter 11 introduces the quantum thermodynamics of heating and cooling and their control.
- Chapter 12 concludes with dynamical control of decoherence.

Part III, consisting of Chapter 13–15, delves into the domain of complex quantum systems and their burgeoning technological applications.

- Chapter 13 introduces quantum tunneling and its information implications.
- Chapter 14 presents the principles of quantum communication, cryptography and teleportation.
- Chapter 15 concludes the book with a brief introduction to quantum computers and the dawn of the quantum information era, commonly referred to as the "second quantum revolution".

May the reader's voyage through the quantum world charted by this book be both pleasant and rewarding.

MATRIX OF CONTENTS

Chapter	Tale (comic strip)	Tale retold	Background	Philosophical issues	Mathematical appendix
Part I Basic Concepts					
1. What is Quantumness?	Henry Bar's debut as quantum superhero.	How small is a quantum?	From atomism to quantum mechanics	From classical to quantum world view: is reality simple or complex?	Constants and variables in quantum physics
2. What is a Quantum Superposition?	Henry Bar splits and recombines	A super-hero in a superposi-tion	Superpositions: from light waves to wavefunctions	Why do scientific paradigms change?	Superposition, wavefunc-tions, vectors and matrices
3. What is Quantum Interference?	The phases of Henry Bar	Henry interferes	Interference in quantum mechanics	The deeper meaning of quantum superposi-tions	Interference and quantum waves
4. What are Quantum Measurements?	The collapse of Henry Bar	Henry is measured	Quantum mechanics as a measurement theory	Parallel evolutions (and universes?)	Projector operators
5. What is Quantum Uncertainty?	Henry Bar grows uncertain	Henry Bar's uncertain position	Uncertainty and complementar-ity in quantum measurement theory	Is uncertainty human?	Continuous variables
6. What is Time–Energy Uncertainty?	Henry Bar's uncertain jumps	The quantum rocket and time–energy uncertainty	The time–energy uncertainty relation in QM	Time and energy in our quantum world	Finite-time evolution

continued ...

Chapter	Tale (comic strip)	Tale retold	Background	Philosophical issues	Mathematical appendix
Part II Quantum Entanglement and Open Quantum Systems					
7. **What is Quantum Entanglement?**	Schred the cat	Schred and Henry are entangled	Entanglement and quantumness	Entangled world	Entangling operators
8. **Entanglement, Decoherence and Which-Path Information**	Schred and Henry disentangle	Decoherence: the dark side of entanglement	Decoherence as which-path distinguishability and entanglement with the environment	On information and free will: do we live in a quantum matrix?	Complementarity between visibility and distinguishability
9. **What is the Environment of Quantum Systems?**	Henry is decohered by the environment	Coherent oscillations and environmental decoherence	Decoherence and decay in an environment (a "bath")	Irreversibility and the arrow of time	Coherent (Rabi) oscillations and decay
10. **Can Quantum Measurements Prevent Change?**	Henry's disrupted wedding	The wedding that never happened	The quantum Zeno and anti-Zeno effects	Is time (or change) an illusion?	Slowing down the evolution
11. **Can Quantum Measurements Control Temperature?**	Henry is scorched and frozen by measurements	Cooling down a heated relationship	Quantum Zeno heating and anti-Zeno cooling	Fiddling with the arrow of time: anomalous thermodynamics	Disrupting the heating process
12. **Can Dephasing be Controlled?**	Henry's dephases and rephases in mid-air	Henry controls his dephased descent	Decoherence and its control	Quantum control of life and death: is change an illusion?	Bang-bang as dephasing control
Part III Quantum Complex Systems and Technologies					
13. **What is Quantum Tunneling?**	Henry goes through walls	Henry challenges impenetrability	Quantum tunneling and wavepacket interference	Motion and its limitations in quantum mechanics	Tunneling and the Schrödinger equation
14. **What is Quantum Teleportation?**	Henry and Schred rescue a distant friend	Teleportation trio	Quantum teleportation and cryptography	Quantum teleportation and transmutation	The quantum teleportation protocol
15. **The Dawn of Quantum Information**	The quantum counter-revolutionaries	Quantum computers: promise and menace	From quantum computing to quantum technology	Long live the quantum revolution!?	A taste of exponential speedup

PART I
Basic concepts

Henry Bar's Debut as Quantum Superhero

CHAPTER 1

What is Quantumness?

1.1 HOW SMALL IS A QUANTUM?

We have found Henry Bar at the turning point of his life, where he is on the verge of becoming the first quantum superhero, having discovered the incredible yet true principle that all things, large and small, are subject to the laws of quantum physics. He finds out that it may be possible, albeit extremely challenging, even for us humans to manifest our "quantumness". This principle underlies Henry's design and implementation of his fabulous quantum suit that allows him to act as a distinctly quantum object in his outlandish adventures.

Henry has made his dramatic discovery as a result of his pondering over deep questions that must be as troubling to the reader as they were to him:

- What is quantum physics?
- What are its laws?
- How do these laws fit into the overall framework of physics and science in general?

Let us recreate Henry's process of coping with these questions by proceeding from the general to the particular issues.

Henry has come to realize that physics is the most comprehensive and fundamental natural science in that it describes and explains the structure and dynamics of all known objects by a set of concise mathematical rules. These rules, when deemed sufficiently general, attain the status of laws of nature.

The diversity of disciplines that have become part of physics is staggering, to Henry's amazement: nowadays, the laws of physics are the key not only to understanding all phenomena that have long been described by branches of physics—e.g., mechanics, electromagnetism and heat—but also to chemistry, biological processes, planetary sciences and cosmology.

He has realized that physics is not only vast but also confidence-inspiring, because since the days of Galileo (in the early seventeenth century) it has followed the arduous yet safe path (at least in hindsight) prescribed by Francis Bacon, the Elizabethan prophet of science:

> **Observation** (*of natural phenomena*) –> **hypothesis** (*concerning their underlying principle or mechanism*) –> **theory** (*culminating in a* **law**) –> **experimental test** *of the theory.*

This path has eventually led to the birth of contemporary physics. Henry's awareness of the fierce scientific battles that marked the evolution of physics has not weakened his conviction that present-day physical theory has been incontestably validated, both by experiments conducted under stringent accuracy bounds and by the scrupulous criteria of mathematical and logical consistency.

Henry, as all of us, has only had first-hand experience of objects behaving according to classical physics. When you push an object, it accelerates and moves along a trajectory determined by the pushing force. Two objects that are so distant that there is no force acting between them behave totally independently.

But quantum objects defy all the foregoing notions: an object can be smeared over a large region of space, and, depending on how you push it, may become either fuzzier (more wavelike) or, conversely, more localized (particle-like). Two quantum objects that were in touch and then receded far apart are still "entangled"; that is, no longer independent. This is only a partial list of strange notions associated with "quantumness".

Now, here is the true spoiler for our recount of Henry's quest: he has come to the conclusion that there is a solid foundation for his astounding finding that **all physics is essentially quantum**. In order to understand and appreciate the boldness of this conclusion we have to pre-empt the historical narrative (Section 1.2) of the emergence of quantum physics in the early twentieth century as a revolutionary theory of radiation, atoms and subatomic particles. It defied the physical theories which reigned supreme at the time: Newton's physics of material bodies and the forces that act between them, or Maxwell's laws of electromagnetic forces and radiation (ranging from radio waves through light waves to gamma rays). Our narrative reveals how quantum physics (or quantum mechanics, as it was originally named) was recognized some ninety years ago to be the only viable theory of atomic and subatomic phenomena and of effects involving very feeble portions of light called photons or quanta. On the other hand, all things ranging from large molecules through living cells or dust particles to people, mountains or planets are still described by "classical"—that is, Newton's—physics.

Henry has become aware of a subtle division between classical and quantum electromagnetic phenomena. Since the mid-nineteenth century it has been known that electromagnetic forces (fields) and radiation—e.g., light—propagate in space as waves. The same is true of photons in quantum physics. The difference is that the energy carried by a single photon, a single "particle of light", cannot be divided into smaller portions, at least not by simple manipulations, so that a photon is a truly indivisible quantum carrier of energy.

For a while, Henry was deeply puzzled by this alleged rift between classical and quantum descriptions: does it mean that there are two (or more) unrelated kinds of physics? If so, should not physics be deprived of its status as a universal "science of everything"? And if, on the contrary, physics is one, how do classical and quantum physics connect?

In the course of his studies, Henry often came across the "correspondence principle": the statement that the application of the rules of quantum physics to objects much heavier or bulkier than an atom yields a "classical" result. Yet this principle did not appear to him compelling: why should the basic rules change as the mass or size grow? Furthermore, in his more advanced reading Henry encountered a remarkable quantum effect known as *superfluidity* whereby a liquid composed of a macroscopic number of helium-3 atoms may flow at very low temperatures as if it were a single quantum object! So, Henry triumphantly concluded that his hunch had been correct: not only mass or size matter when discerning a quantum object from a classical one. But then how to decide whether an object is quantum? And is such a decision always clear-cut?

Having acquired a deeper understanding of quantum physics, it has become clear to Henry that the key quantity in assessing quantumness is "**action**": the change in the energy of the object multiplied by the duration of the change. The essence of quantumness in nature is that action cannot be less than an elementary portion; a quantum in Latin. How small is this quantum of action? As explained in the appendix (Section 1.4), it is an exceedingly small number, common to all processes in nature (a universal constant). It is known as Planck's constant, which is denoted by the symbol $\hbar$. Henry Bar's name is a pun on this symbol, which is the "coat of arms" of our quantum superhero.

In order to appreciate the smallness of $\hbar$, Henry analyzed the action of a very lightweight and compact object: a microscopic pendulum which weighs one billionth of a gram and is attached to a micrometer-long cord. This calculation shows that a kick that would set this pendulum in motion would still be 1,000,000,000,000,000,000 times larger than $\hbar$; to observe a single quantum of action we would then need an accuracy of 1 part in 10^{18}!

This analysis has been an eye-opener for Henry: he has realized that sufficiently high accuracy of measurements or control can make any object, no matter how heavy or large, reveal its quantumness! This conclusion has been reinforced by Henry's further reading that nowadays quantum effects are being measured for clouds of millions of ultracold atoms or nanomechanical cantilevers and membranes, but as late as the 1990s such ultraprecise measurements would have been considered a pipedream (Chapters 2–4). The size, weight and complexity of objects that act quantum-mechanically keeps growing by the year. That is why experiments probing the quantumness of man-size objects such as Henry, although unfeasible today, cannot be ruled out in the future.

Henry has thus come to the conclusion that the boundary between quantum and classical objects is largely arbitrary. Current experiments are already capable of revealing the quantumness of objects visible to the naked eye; techniques whose germs already exist, as we will show, can push this boundary to the extent that even macroscopic objects may display quantum behavior.

Notwithstanding the formidable technical challenges that may prevent us from observing quantum features of various objects, a much more general insight transpires from the foregoing discussion: *the universality of quantum description*. Quantumness is lurking beneath classical phenomena, **physics is one**, and it is up to us to reveal its quantum face, should we wish so. As the Henry Odyssey unfolds, so will the narrative of the effects of quantum physics, from the simplest to the more advanced.

1.2 FROM ATOMISM TO QUANTUM MECHANICS

Let us trace the origins of Henry's initial view shared by many scientists and the lay public alike, which draws a line, crudely speaking, between the quantum description that is incontestable in explaining atomic or subatomic phenomena and the classical (Newton's, Einstein's, or Maxwell's, as the case may be) description of the macroscopic world. How did this view come about?

Its roots are in the notion of atomism that originated in ancient Greece and took more than two millennia to become a universal theory of the "true" reality beneath the realm of our everyday (macroscopic) experience. The evolution of atomism and the emergence of quantum physics from these roots will be sketched in what follows.

a) *The atom makes its appearance*. The idea of an atom as the ultimate, indivisible (a-tom in Greek) and immutable constituent of matter is attributed to Democritus (Figure 1.1), "the laughing philosopher", in the fifth century BC. This idea won many adepts in the Greco-Roman world among followers of Epicurean and Stoic philosophy. They were drawn by the blind, purposeless world view of atomism: atoms collide at random, combine to form an object, then randomly recede and disintegrate the object, then recombine, and so on and so forth, forever. Its opponents loathed this world view, either on religious grounds or because it defied their logic. It is unfortunate for the development of science that Aristotle in the fourth century BC ridiculed atomism because it contradicted his notion that the properties of objects are immutable, as opposed to the atomistic view that all objects incessantly change, disappear and are born again through atomic collisions and only the atoms are unchanged. The indisputable authority of Aristotle condemned atomism to oblivion. Still, this idea resurfaced in the late seventeenth century and acquired repute in the nineteenth century.

One discipline where atomism found a fertile ground was chemistry in the aftermath of the revolutionary discovery by the French lawyer-turned-scientist A. Lavoisier (shortly before his head was chopped off by a revolutionary tribunal in 1792) that in chemical reactions the mass (weight) ratios between the reacting substances are fixed. The British scientist J. Dalton (Figure 1.1) surmised in 1803 that Lavoisier's mysterious discovery could be explained by the supposition that all substances are composed of atoms with fixed specific masses. The Italian scientist A. Avogadro proposed in 1811 that vessels of the same volume should contain the same number of atoms regardless of the atomic mass, if we assume that the atoms are tightly packed in the vessel as tiny balls. A hundred years later this universal number was quantified, measured and named after Avogadro. Perhaps the greatest achievement of atomism in nineteenth-century

Fig. 1.1 The Fathers of atomism—Democritus (left) and Dalton (right)—and their ideas. Democritus posited that atoms of different shape, size and smoothness combine to form different substances. Dalton assumed that atoms bind to form molecules.

chemistry was the arrangement of all chemical elements in a periodic table by the Russian chemist D. Mendeleev, who arranged the elements in the table according to a mysterious index which he named the "atomic number". It was not until the quantum theory of atoms emerged in the twentieth century that the atomic number was understood to be determined by the inner structure of the atom, which had been completely unknown at the time of Mendeleev! Despite these tremendous nineteenth-century discoveries that relied on atomism, in the absence of direct evidence for the existence of atoms, their opponents remained unconvinced, as our tale will show.

Another discipline where a handful of bold scientists resorted to atomism was the physics of gases. In 1738 the Italian scientist D. Bernoulli gave an atomistic explanation to the seventeenth-century R. Boyle's law whereby gas pressure grows inversely with the volume of the container. According to Bernoulli, pressure was a force exerted on the container walls by colliding gas atoms, and Newtonian mechanics implied that this force would increase as the container became tighter. Unfortunately, Newton's own opposition to atomism hindered the progress of this idea. Oddly enough, Newton believed in the corpuscular (particle-like) nature of light, but not of matter.

It was the Scottish physicist J. C. Maxwell (Figure 1.2) who gave atomism a boost in the mid-nineteenth century by introducing the concept of random collisions among atoms in a gas: he maintained that the average velocity of many colliding atoms determines the gas temperature, but an individual atom may have velocity, of which we may know only the statistical probability. This atomic theory of gases was embraced and further developed by the Austrian physicist L. E. Boltzmann (Figure 1.2) and the American scientist J. W. Gibbs in the late nineteenth century. Yet it was met by vehement opposition. The influential physicist and philosopher from Prague, E. Mach, objected to atoms on "positivist" grounds: if you cannot see or detect an atom and cannot even ascribe a definite velocity or path to it, then it is fiction that does not represent "positive" data. F. W. Ostwald, the German founder of physical chemistry, was another objector to atoms on similar grounds, but mainly because he had an alternative theory of matter. Most of the attacks on atomism were directed against Boltzmann, who viewed the random, statistical description of atoms as the basis of heat exchange theory known as thermodynamics. Such statistical description was considered by many to be an abomination to the exact science of certainty that physics, including thermodynamics, was supposed to be. In the face of so much opposition, Boltzmann, who felt that his life's work had been rejected, killed himself in one of his depressive moods in 1905.

Fig. 1.2 Maxwell (left) and Boltzmann (right). The greatest nineteenth-century atomists posited that random collisions among many atoms in a vessel are the origin of thermodynamic laws.

Little did Boltzmann know that, within months of his death, Einstein's theory of the random motion of a macroscopic particle in a liquid (known as Brownian motion) would fully vindicate the reality of the atomic (or molecular) composition of liquids, and suggest a way of inferring the atoms' presence and number (Avogadro's number) from the rate at which the macroscopic particle diffuses in the liquid under random collisions by the atoms. A few years later, the influential French physicist J. Perrin concluded, on the basis of all available evidence, that atoms were no longer a hypothesis but scientific fact: the atom finally prevailed! The ancient atomists would have probably been thrilled at this acceptance of their notions of random collisions between indivisible entities as the key to observable phenomena. Mach, on the other hand, persisted in his denial of the existence of atoms till his death. Evidently, scientists may also have their biases.

b) *The atom splits.* Just as the status of the atom was elevated from fiction to fact, another conceptual change came about. Two landmark experiments convincingly demonstrated not just the existence of atoms but also their divisibility that contradicted their very name and original notion. J. J. Thomson (Britain) measured the fragments of gas atoms shattered by electric impulses. Upon applying an electric force to the gas, Thomson measured the curvature of the fragment trajectories and deduced from Newtonian mechanics the ratio of the electric charge to the mass of the fragments. One fragment, nicknamed the "nucleus", was found to be positively charged and thousands of times heavier than the other, negatively charged fragment, termed the *electron*. Subsequently, E. Rutherford (New Zealand/Britain) (Figure 1.3) measured by similar methods the charges and masses of fragments of the atomic nucleus that disintegrated by radioactive decay. The atom was now real and divisible.

Rutherford thought of the atom as a miniature planetary system: a heavy nucleus encircled by orbiting, much lighter electrons. Yet it was unclear what kept only certain electronic orbits stable, but not others. Data on radiation emission and absorption by atoms suggested that electrons in atoms changed their orbital motion in a peculiar, inexplicable manner. A new theory was acutely needed.

c) *Radiation goes quantum*. In 1900 the German physicist M. Planck (Figure 1.4) presented his astounding theory of the properties of radiation emanating from an almost closed box or cavity (oven). Planck's theory was meant to explain the experimentally observed dependence of the radiation frequency on the temperature of the oven (cavity) walls. Planck came to the momentous conclusion that no equilibrium between the absorption and emission of radiation by the oven walls is possible unless we suppose that the radiation can take up or give off energy only in tiny, discrete portions (quanta in Latin). There was no analog to this radical conclusion in the physics of the day, namely, in thermodynamics or in Maxwell's theory of electromagnetic radiation wherein radiation energy at a given frequency may have arbitrary, not just discrete, values. However, Planck noted that without the discreteness (quantum) assumption the existing theory would lead to the absurd conclusion that the total radiated intensity was infinite, which meant that radiation could not be in equilibrium with the walls, contrary

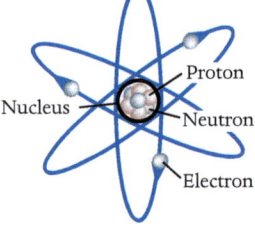

Fig. 1.3 Rutherford—the discoverer of atom splitting—and his planetary model of the atom

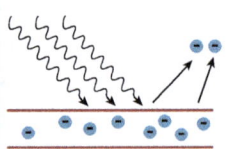

Fig. 1.4 The Fathers of radiation quanta—Planck (left) and Einstein (right)—and the photoelectric effect whereby light quanta knock out electrons from a metal.

to the experimental evidence! Yet Planck was reluctant to think of quanta as real objects. After years of attempts to reconcile this notion with Maxwell's electromagnetism and thermodynamics, to no avail, he was forced to recognize that, unintentionally, he had opened up a new era in physics: *quantum physics was born in 1900!*

d) *Quanta become real.* Five years later, in 1905, A. Einstein (Figure 1.4) took the notion of light quanta in earnest in his theory of P. Lenard's photoelectric effect; namely, the ejection of electrons from a metal surface by light. Einstein asserted that light quanta are essential for understanding the troubling experimental finding that the velocities of the ejected electrons are unaffected by the light intensity. This finding contradicted Maxwell's theory whereby light intensity corresponds to the force that accelerates the electrons and thus determines their velocities. Einstein's explanation of this contradiction was that indeed Maxwell's theory is inadequate here. Instead, one should take into account that a quantum of light must have sufficient energy, which is proportional to its frequency, to release an electron from its "trap" formed by the surrounding charges in the metal (the "work function"). There is thus a sharp threshold in the energy or frequency of the quanta above which electrons are ejected. However, an increase in the light intensity at a frequency above this threshold would merely increase the number of light quanta and therefore the number of ejected electrons but not their velocities, in agreement with experiment. Einstein's "exotic" explanation of the photoelectric effect, for which he was later awarded the Nobel Prize, convinced many physicists that the reality of quanta was plausible. Einstein's view was that light quanta (termed *photons*) represented the "granularity of the electromagnetic field in space and time". This granularity reminded Newton's corpuscular theory of light that was defeated by wave optics and Maxwell's theory of electromagnetic waves in the nineteenth century (see Section 2.2). Yet, photons were much more puzzling: they appeared to be both particles and waves! This bizarre duality called for further explanation.

e) *The atom goes quantum.* In 1913 N. Bohr (Denmark) (Figure 1.5) put forward a quantum model of the atom, which, unlike the earlier planetary model by Rutherford, explained the observed frequencies (or wavelengths) of radiation absorbed and emitted by atoms. The word "quantum" in this model, inspired by Einstein and Planck, referred to the discrete values of these frequencies ("spectral lines"). The great achievement of Bohr's quantum model was its ability to explain the puzzling *discreteness* of the solar spectrum; namely, the frequencies of the lines associated with light emission from hydrogen gas in the sun.

Implicit in Bohr's model (later expounded by A. Sommerfeld, Germany) was the idea that the electron bound to the atomic nucleus behaves as a wave: it circles

Fig. 1.5 Prince de Broglie and his matter waves (left); Bohr and his quantized (standing wave) electron orbitals in an atom (right).

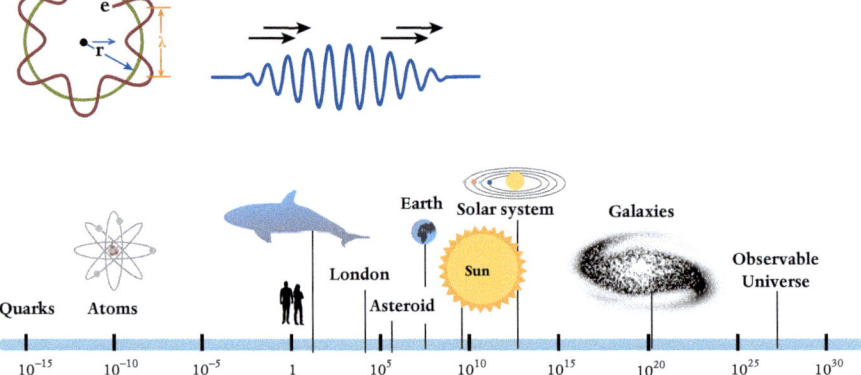

Fig. 1.6 Size scales (in cm) from the observable universe down to sub-nuclear particles (quarks).

around the nucleus in an orbit only if it forms a standing wave (Section 1.3). Hence, its energy is discrete (quantized) because successively larger orbits ("orbitals") are realized by standing waves that correspond to successive multiples ("levels") of the lowest orbital energy. Thus, the notion of quantum waves was introduced into theoretical physics.

The notion of quantum waves of matter, on which we shall elaborate in Chapter 2, quickly became the cornerstone of a new, universal description of microscopic objects, nicknamed "quantum mechanics" (QM). Its universality was based on the striking conclusion shared by Planck, Einstein, Bohr and de Broglie (Figure 1.5) that all quanta of energy, be it in light, atoms or free electrons, are multiples of the same tiny number: Planck's constant, denoted by $\hbar$ (see Section 1.4, Appendix). The smallness of this constant appeared to delineate the boundary between microscopic scales describable by QM and macroscopic scales ruled by classical mechanics and electromagnetism (Figure 1.6). However, a much clearer distinction between classical and quantum phenomena emerged after 1926 (Chapters 2–4).

1.3 FROM CLASSICAL TO QUANTUM WORLD VIEW: IS REALITY SIMPLE OR COMPLEX?

The inception of quantum physics or quantum mechanics (QM) was a landmark in the age-old quest for an answer to a key question of natural philosophy: is reality, although manifest by an endless variety of seemingly complex phenomena, reducible to a few simple constituents?

The human strive for simplicity was expressed with extreme poignancy in an aphorism of the first Greek philosopher Thales (seventh century BC): "Everything is water". Crude as it may appear, this aphorism expresses the first known attempt to construct a "theory of everything".

More elaborate but still simple was the idea of atomism introduced by Democritus (Section 1.2), who summarized it thus: "According to the senses—there are color, taste, smell; but according to reason—only atoms and void." The known quantum physicist Feynman expressed his admiration for the idea of atomism thus: "If . . . all of scientific knowledge were to be destroyed, and only one sentence passed on to the next generations of creatures . . . it is . . . that all things are made of atoms".

Such reductionism of the complex reality to simple constituents characterized nineteenth-century atomism in chemistry and gas theory. Despite the discovery that the atom has structure, as was revealed by Thomson's and Rutherford's experiments, scientists insisted on reductionist simplicity in their thinking of the atom. This is evident in Rutherford's (planetary) model of the atom.

Early quantum theory by Planck, Einstein, Bohr and de Broglie pushed reductionism even further by introducing elementary units (quanta) of energy and thus reinforced the view of universal simplicity of atoms and their constituents.

Yet subsequent developments in QM have led to a revision of this view and culminated in a more elaborate scheme of reality, where many attributes are needed to classify the basic properties of atomic, subatomic or sub-nuclear particles that have by now become highly complex entities. Furthermore, atoms and their constituents are, in general, no longer immutable or completely stable: atoms may undergo transmutation and their electrons change their energies in an elaborate way. Subatomic particles are subject to even more drastic changes, and, with very few exceptions (such as the proton and the electron), do not last forever.

Not only do these modern notions undermine the tenets of original atomism, they also pose a serious challenge: can we really describe phenomena occurring on macroscopic scales, including material properties, by means of these not quite simple "building blocks" of matter that abide by QM? Even if conceptually this is the case, it can hardly be put to a practical test, as any attempt to use QM tools to calculate or predict macroscopic phenomena is doomed to fail: even the largest computing resources imaginable cannot cope with the complexity of calculating the evolution of large numbers of interacting atoms quantum-mechanically. Therefore, molecules comprised of tens, let alone hundreds, of atoms cannot be exactly analyzed by QM. Instead, M. Levitt, A. Warshel (Israel, later the USA) and M. Karplus (USA), who were awarded the Nobel Prize in Chemistry in 2013, combine QM and classical methods to calculate the structure and dynamics of such multi-atomic molecules.

At present, the only way to describe the intricate details of systems composed of more than a few atoms is to adopt approximate methods tailored to the level of complexity of the system at hand. This means that we effectively introduce *different physical rules* for different levels of complexity: this is the essence of modern chemistry, condensed-matter physics and other forms of many-body physics. Even more extreme are the approximations that underlie statistical physics or thermodynamics, which are powerful tools for predicting the average evolution of large collections of atoms but tell us practically nothing about individual atoms.

The moral is that constructing a description of highly complex systems from that of its elementary constituents may well be practically and even principally prohibitive. To stress this point, the term *emergent properties* has been coined: it denotes characteristics of complex systems that cannot be straightforwardly inferred from those of their ingredients. Reality may be conceptually simple, but this simplicity is only manifest when we isolate a few atoms, photons or subatomic particles from the rest of the world. S. Haroche (France) and D. Wineland (USA) were awarded the Nobel Prize in Physics in 2012 for having pursued this approach, which reveals QM in all its glory. A major challenge for science in the twenty-first century is to extend calculational and experimental QM tools to macroscopic systems, thereby merging the "bottom-up" (progressing from simple to complex phenomena) and "top-down" (progressing conversely) approaches to understanding reality.

Yet there is a deeper unresolved issue: how far up the complexity ladder can we push QM as a conceptual framework for explaining reality? In particular, is QM at all relevant to biological processes? There are currently many proponents of a discipline that has been termed *quantum biology*, but the evidence for truly quantum effects in the functioning of live organisms is still too flimsy to pass judgment on its validity. Even more far-fetched are the recently proposed applications of QM notions to the domains of human consciousness, psychology and social structure. However, those involved in such endeavors are not trying to unify physics and human sciences in a common framework, which may be futile (or not?), but rather borrow quantum tools for the construction of models that bear analogy to QM for whatever reason. QM is just too powerful a trick to pass by!

A Fuzzy World

When contemplating our world
The gist of it you should behold
And reach the strange but clear conclusion:
Its ruggedness is an illusion.
Its stuff is but a fuzzy cloud
Uncertain, flimsy and spread out.
But when perceived, it may be asked:
Can its wave-nature be unmasked?

1.4 APPENDIX: CONSTANTS AND VARIABLES IN QUANTUM PHYSICS

This appendix introduces the basic mathematical notions and notations that will serve in subsequent appendices to describe features and phenomena of quantum physics.

Mathematical methods of physics in general employ constants and variables. Constants are numbers that quantify unchanging physical relations. Usually, a constant is denoted by a designated letter or symbol, to avoid the need of repeating the entire number every time. Constants play such an important role in physics that they are the subject of a central field of research known as *metrology*. Numerous physicists in this field perform elaborate experiments trying to ascertain the *exact* value of these constants.

Let us consider several fundamental constants. The *speed of light* is one such constant. Its symbol is the letter c and its value is 299,792,458 meters per second. This means that light travels around 300,000 kilometers in one second. Thus, for example, it takes around 8 minutes for light from the sun to reach Earth.

The fundamental constant of quantum physics, and Henry Bar's superhero symbol, is $\hbar$, (hence the name "Henry Bar"). This constant is known as *the reduced Planck's constant*, where Planck's constant is designated simply by the letter h (without the bar) and the two are related by $\hbar = h/2\pi$. Why is Planck's constant so important that we named our quantum superhero after it? The full answer requires reading a good deal more of this book. Here, suffice it to say that h relates the frequency of a quantum of light to its energy: $E = hf$, where E is the energy and f is the frequency of the quantum. This means that h is the change in energy (whose unit is called Joule, designated J) produced by a change in frequency (whose unit is called Hertz, designated Hz) of the quantum. In other words, if a quantum of light increases its frequency by 1 Hz, its energy is increased by h J.

The value of Planck's reduced constant is $\hbar = 1.054 \ 10^{-34}$ J/Hz. The meaning of 10^{-34} is 1 divided by 100,000,000,000,000,000,000,000,000,000,000,000. This is an *extremely* small number. The common notion that "quantumness is only manifest for small things" stems from Planck's constant, which appears in almost any formula in quantum physics, being so small. However, throughout this book we will come across phenomena that defy this notion, as does Henry.

A much more common type of mathematical objects that we shall encounter are letters and symbols that represent quantities that change; that is, they have *variable* values. An example is the position of an object, usually denoted by the letter x, that can acquire different values as the object moves. In the explanation of $\hbar$ above, E and f are variables denoting energy and frequency, respectively, that can change due to a dynamical process.

A ubiquitous mathematical notation used in physics is that of indices, which are usually designated by either superscripts or subscripts. For example, x_1 may denote the position of one thing and x_2 the position of another. If we have one hundred objects, labeling them x_1, x_2, x_3 ... up to x_{100} would be very cumbersome. Instead, indices obey the following notation: x_i ($i = 1, \ldots, 100$), which means that x_i represent the positions of all objects, from 1 to 100.

Let us introduce another extremely important variable, time, denoted by the letter t. In physics, time is measured in seconds and plays a pivotal role in describing change. Thus, for example, speed is the *change in position over the time interval* that this change takes, $v = (x_2-x_1)/(t_2-t_1)$, where x_1 and x_2 denote the position of the same object at two times, t_1 and t_2, respectively. We will introduce a more sophisticated way to represent changes, namely derivatives, in subsequent chapters.

Another common notation is that of summation. Say we have one hundred objects with different masses, measured in grams, g. The notation $M = m_1+m_2 + \ldots +m_{100}$ for the sum of their masses is too cumbersome. Instead, the compact notation of a sum sign, $\sum$, can be introduced, which allows us to write: $M = \sum_{i=1}^{100} m_i$. Below the sum sign we have $i = 1$, which denotes the first value of the index of summation, here i. Above the sum sign we have 100, the last value of the summation index. After the sum sign comes the indexed variable that is the object of summation; here the mass, m_i.

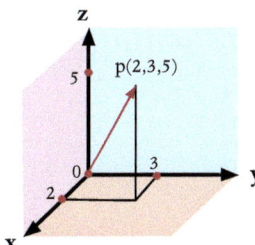

Fig. 1.7 Position along mutually perpendicular axes denoted by a three-dimensional vector.

As a more complex example, consider a beam of light made of light quanta, called photons, each with its own frequency. We have f_i $i = 1, \ldots, 10^3$, denoting the frequencies of all light quanta. Their *total energy*, as explained above, is $E_{total} = \sum_{i=1}^{10^3} E_i = \sum_{i=1}^{10^3} hf_i = h \sum_{i=1}^{10^3} f_i$. The first equality uses the summation symbol over the energies of all the quanta; the second equality uses the above relation between energy and frequency *for each light quantum*; and the last equality uses the fact that Planck's constant is the same for all quanta to rewrite the total sum as Planck's constant times the sum of the light-quanta frequencies.

The last notation we wish to introduce is that of a *vector*: a collection of variables describing a specific entity. As an example we consider the three-dimensional (3D) position of an object; namely, its position along the mutually perpendicular x-, y- and z-axes. To be specific, let us take the x-axis to extend from the backward to the forward direction, the y-axis from the left to the right, and the z-axis to point from the bottom to the top (Figure 1.7). Instead of specifying an object's position by the variables x,y,z, we denote it by the vector $\underline{x} = (x,y,z)$.

One may perform simple operations on vectors. For example, vector addition, $\underline{x_3} = \underline{x_1} + \underline{x_2}$ signifies a vector that has the form $\underline{x_3} = (x_1+x_2, y_1+y_2, z_1+z_2)$, wherein the values in each dimension are summed independently. The same applies to subtraction. Vector multiplication, denoted by a dot-product, $\underline{x_3} = \underline{x_1} * \cdot \underline{x_2}$, obeys the rule $\underline{x_3} = (x_1 * x_2, y_1 * y_2, z_1 * z_2)$.

To summarize, in this appendix we have introduced the notion of constants, dwelling on the speed of light and Planck's constant, and the notion of variables. The latter can be indexed to account for multiple objects or components and can be "grouped" into vectors, for which the summation notation and vector operations, such as summation and dot-product (vector multiplication) were defined. In subsequent chapters we will use these notations to describe some quantum-physical phenomena that Henry Bar encounters.

Henry Bar Splits and Recombines

CHAPTER 2

What is a Quantum Superposition?

2.1 A SUPERHERO IN A SUPERPOSITION

Henry has just become the first quantum superhero, having overcome the tremendous technological obstacles en route to the implementation of his quantum suit. In the previous episode he successfully tested the first function of his quantum suit, the Split button. Congratulations are in order, as he has thereby freed himself and mankind from the constraints of our mundane "classical" existence by revealing our latent quantum features.

The feature that the Split button exposes is the ability of a quantum object to be *in several places at the same time*. More precisely, it is the ability to propagate along different paths simultaneously, which Henry experienced by walking through both the revolving door and the sliding door at once. How can such a bizarre phenomenon occur?

It certainly contradicts the behavior of a material object that is ruled by Newton's "classical" physics. Even if the object is extremely small, such as a typical molecule, it is expected to maintain its cohesiveness at all times, so that it does not blur or diffuse as it moves along a (unique) trajectory determined by the forces that act upon it. At each time instant, the position of the object is specified by a set of numbers—its coordinates. Instead, Henry's Split button replaces his instantaneous position by two positions, each specified by different coordinates—a so-called *"superposition"*.

This behavior appears less bizarre if we think of Henry as a *wavelike* object. Consider a wave in a pond. It does not have a single position but instead is "smeared" over many positions throughout the pond at each time instant. Accordingly, it is described by a large set of coordinates simultaneously. As it propagates in the pond, it spreads out more and more.

Such wavelike properties have long been known in branches of "classical" physics that have evolved from Newton's mechanics: acoustics (the physics of sound waves in gases), hydrodynamics (the physics of waves in liquids), and in classical optics which is ruled by Maxwell's electromagnetic theory.

Yet Henry is not merely a wave either. He is aware of being a *single entity*—one Henry, even when he splits. He becomes a superposition of two lookalikes (versions), but they are not his clones or copies, as they share the mass of the original Henry (mass conservation is undisputed in this case). More strangely, as we shall elaborate in Chapter 4, any attempt to localize (pin down) one of the lookalikes will recreate the whole Henry again, not a fraction of him. The lookalikes only tell us where Henry can be found if localized, not where he actually is. *He can be anywhere before he localizes.*

This strange property distinguishes a quantum wave from the classical waves mentioned previously. As discussed in Chapter 1, quantum physics originated from the concept of light quanta, which propagate as waves, but maintain the indivisibility of their energy: a single photon cannot be divided by a splitting operation into half-photons that may emerge at different points of space and time, as we shall always find either one or zero photons. The same is true for electrons or other subatomic particles, as discussed in Section 2.2. Half-particles cannot be found, only whole particles or none at all. Counterintuitive as these properties may appear, they have been confirmed by a multitude of experiments over the past century.

This unique character of quantum waves is captured by a mathematical description known as a "wavefunction"—a set of numbers that are assigned to every position (a point in space) at a given time. These numbers are known as the "probability amplitude" of the wave at that point. In Henry's case his two lookalikes represent equal probability amplitudes of finding him at the revolving door or the sliding door, so that his chances (probability) of emerging (i.e., being localized) here or there are 50%/50%. Yet before he localizes (materializes) *we cannot tell where he is.*

The splitting phenomena described above are not restricted to a superposition of two quantum waves (or probability amplitudes), but actually apply to superpositions of any number of such waves. During Henry's ride through town, he pushes his Split button many times, each time becoming more spread out, until the streets are flooded with many of Henry's lookalikes. The probability amplitude of each lookalike is then small, and the probability (which is the probability amplitude squared) of finding him in any given street is still much smaller.

But why is Henry's splitting so outlandish that it can only be revealed by a futuristic contraption? Let us keep in mind that such splitting requires the object to become a superposition of wavelike components (probability amplitudes) that are spatially *distinguishable,* but partly overlap to ensure they are mutually *coherent,* i.e. stem from one wave—such as Henry's wavelike lookalikes that ride along different streets or pass through different doors, but still maintain their coherence ("oneness"). Both requirements are extremely harsh to fulfill by an energetic, massive object. The reason is that the wavelike "fuzziness" or smearing of a quantum object becomes smaller as its mass and energy grow, so that experiments capable of revealing this property become harder.

Let us make this discussion somewhat quantitative. As noted in Section 2.2, an electron with an energy obtained by running it through a tension of 100 Volts was first shown to be a wave by scattering it off atoms in a crystal that are 0.4 nanometers apart, giving rise to a superposition of scattered electron waves separated by roughly the extent of one such wave. Had the distance between atoms in the crystal been much larger, the scattered electron waves would not have exhibited coherence, i.e. they would not have "added up" (interfered) at the detector, as explained in Chapter 3. Much heavier objects, e.g. atoms or molecules, must be endowed with much less motional (kinetic) energy in order to observe their coherent splitting into wavelike components.

It then transpires that Henry's mass being so large, he must be given an exceedingly small kinetic energy to allow for his coherent splitting on the scale portrayed in our cartoons. Giving or receiving such tiny energy is an incredible challenge at present, but one which does not violate the known laws of physics.

Henry's quantum suit has been equipped not only with the Split function but also with its reverse: the Recombine function that reunites Henry's lookalikes into his original self. Such functions are not only figments of our imagination; they are readily available experimentally. For example, consider an electron bound to an atomic nucleus (Figure 2.1). Initially, the electron wavefunction has a well-defined amount ("level") of energy, as it (crudely speaking) occupies an orbital that encircles the nucleus. Now we shine a laser pulse at the atom. If we choose the pulse to have the right intensity, frequency and duration it will cause the electron wavefunction to split into an equal superposition of two wavefunctions corresponding to different energy levels or orbitals. An identical subsequent laser pulse will recombine the superposition and restore the electron wavefunction to its original form. This example, which will be revisited in Chapter 3, illustrates our ability to do and undo splitting quantum operations of the kind that Henry performed on himself.

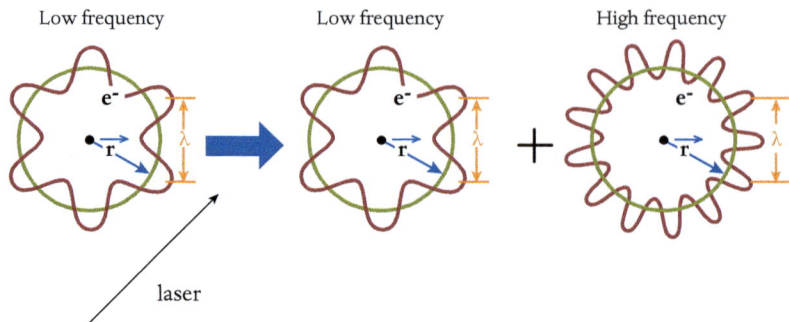

Fig. 2.1 Laser-induced splitting of an atomic-electron low-energy (or low-frequency) wavefunction into a superposition of energy levels (orbitals corresponding to low and high frequency or energy, respectively).

A superposition of the atom's electron orbitals is an example of a superposed internal degree of freedom; namely, the atom may propagate as a single object through space while its internal state is described by a superposition of two or more wavefunctions with different energies.

As our story unfolds, Henry gradually uncovers his quantum potentialities and takes advantage of them by adding more functionalities to his quantum suit, and taking advantage of other wavelike properties. He can now revise the battle song of the *"Scarlet Pimpernel"*:

The elusive Henry Bar

She seeks him here, she seeks him there—
His foe—she seeks him everywhere.
Is he at the sliding or the revolving door?
At both! Such is the quantum-wave galore!

2.2 SUPERPOSITIONS: FROM LIGHT WAVES TO WAVEFUNCTIONS

In order to understand in depth Henry Bar's quantum wavelike behavior, let us review the emergence of the relevant concepts and their subsequent development which has crystallized quantum physics. Strikingly, the decisive stage in the formation of this revolutionary theory that rules physics to this day occurred in a span of few years, during the true heyday of quantum physics: 1923–27. However, it had taken more than a century for this revolution to come to fruition.

a) *Light waves emerge.* That light propagates as a wave was convincingly argued by the Dutch scientist C. Huygens (in the mid-seventeenth century), as opposed to Newton, who maintained that light consists of particles (corpuscles). Whereas Newton's corpuscles / particles propagate along straight lines known as rays, light waves, according to Huygens, scatter from obstacles or emanate from sources as expanding spheres. Huygens had no direct evidence for his theory of light, which was rejected and remained dormant for some 150 years. An experiment of the British physicist T. Young in London in 1805 demonstrated the ability of light to penetrate two slits in a correlated (coherent) fashion called "interference" (Chapter 3). A. J. Fresnel's experiments (c.1830) in France displayed wavelike light penetration through pinholes—an effect called "diffraction"—and revealed the conditions under which light propagates as concentric, spherical waves emanating from a point source, rather than straight-line rays. All of Fresnel's effects could be inferred from the comprehensive theory (developed in the 1860s) of the Scottish physicist J. C. Maxwell, which treated light as an electromagnetic wave; i.e., as a propagation of electric and magnetic fields that alternate in space and time. Maxwell's theory of electromagnetism was quickly recognized to be the pinnacle of our understanding of all forms of electromagnetic radiation, and it is still widely used today. Yet at the time it triggered a heated debate as to whether electromagnetic waves propagate in empty space or in a hypothetical medium called "ether". The ether theory was dispelled with an experiment by the American physicist A. A. Michelson (who nevertheless continued to believe in it). This experiment confirmed the special relativity theory of the Dutch physicist H. A. Lorentz and of Einstein (1905), whereby light-wave velocity is constant; i.e., independent of the light-source motion. Such motion would have affected light-velocity had light propagated in ether, depending on whether the light source (affixed to the Earth's surface) had receded from this ether or approached it.

b) *Quanta emerge.* The next revolution was the concept of light quanta introduced by Planck and Einstein (reviewed in Chapter 1). Remarkably, already in 1909, G. J. Taylor (Britain) experimentally tested the propagation properties of a single photon and came to the same conclusion as Einstein: that a photon propagates as a wave. His ingenious experiment consisted in the propagation of exceedingly feeble light—approximately one photon—through a pinhole, and photographing the resulting spatial pattern which consisted of concentric rings, just like Fresnel's pattern. To be able to record the effect of such a tiny amount of light on a photographic plate, Taylor had installed a slow-burning candle inside a photographic dark chamber, sealed the setup, placed a "do not touch" sign on top, and in the fashion of a gentleman-scientist of the day, went yachting for more than three months(!), which was the necessary exposure time for photographing a *single* photon. The recorded pattern (Figure 2.2) indicated that a single photon indeed propagates as a wave!

c) *From quantum waves to wave mechanics*. In 1923 a Parisian PhD student, Prince Louis de Broglie, published his revolutionary surmise that a material object, such as an electron, moves as a packet (superposition) of waves with different energies, alias a "wavepacket" (Figure 2.3). This meant that much like light waves, "matter waves" are delocalized—spread throughout space—so that they may be found at different locations at the same time.

In essence, de Broglie extended the notion of a quantum, introduced into electromagnetic theory by Planck and Einstein, to a matter wave—i.e. a massive object—following the hint contained in Bohr's theory of the atom, wherein the

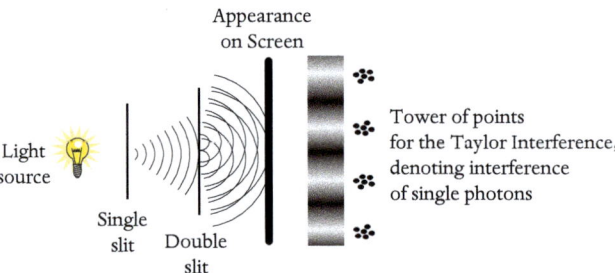

Fig. 2.2 Fresnel's pinhole diffraction (light passing one slit), Young's two-slit interference (periodic bright and dark stripes on the screen) and Taylor's single-quantum interference (periodic dot pattern on the screen, each dot representing a quantum or photon that strikes the screen).

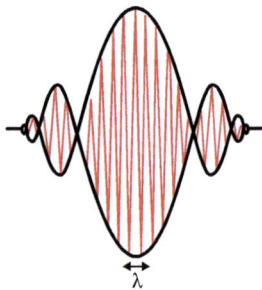

Fig. 2.3 A wavepacket describing the motion of a quantum particle governed by the Schrödinger equation. The amplitude of the interfering or superposed quantum waves forming the wavepacket oscillates in space on the scale of the de Broglie wavelength λ. The dotted contour denotes the amplitude variation in space between crests and troughs on much larger scales. This contour indicates the extent to which the position of a particle is smeared by its quantumness.

atomic electron forms a standing wave in any of its admissible orbits (orbitals) around the nucleus (Chapter 1). Yet the quantum wave of de Broglie has the unprecedented property that although it can be split between two slits, it still retains its "oneness"; i.e., one cannot find half the quantum (or half the electron) in each slit, as opposed to a classical Maxwell's wave.

Another major difference between de Broglie waves and classical electromagnetic waves is that the scale of wavelike properties (wavelength) of a material particle diminishes as its mass grows. Therefore, the de Broglie wavelength is exceedingly tiny for macroscopic objects (Chapter 1).

d) *Wave mechanics is born*. The essential differences between de Broglie's quantum matter waves and classical light waves prompted the Viennese physicist E. Schrödinger to work out, in 1926, what has since become the cornerstone of quantum mechanics (QM): the equation of motion for matter waves, alias the wave equation. The strength of the wave equation is its generality: it applies to any material object, be it a free-propagating particle, such as an electron scattered by atoms, or a particle confined by a potential, such as an electron bound to an atomic nucleus.

The ability to describe material particles as waves, which at the time was known as "wave mechanics", was experimentally confirmed a year later (1927) by C. Davisson and L. Germer (in the USA) and G. Thomson (in the UK) (Figure 2.4). They observed that an electron is simultaneously scattered by several atoms in a crystal just as a quantum wave would be, while retaining its "oneness" (coherence). Schrödinger's wave equation now had its crucial proof.

In the course of few years, the Schrödinger wave equation was recognized to be universal and fundamental; i.e., applicable to any physical system and any observable quantity. The full significance of this statement will be gradually clarified throughout this book. But what does this equation actually tell us?

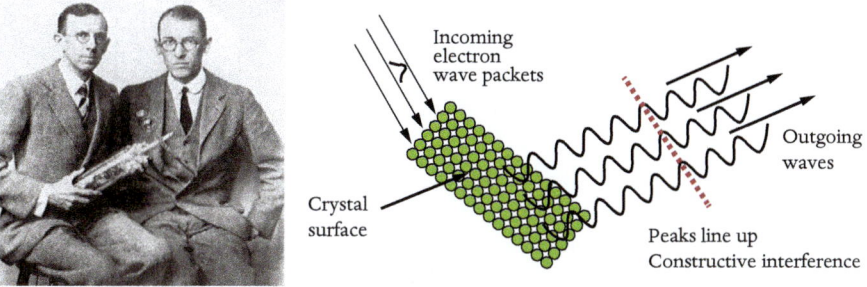

Fig. 2.4 Davisson (left) and Germer (right) and their demonstration of electron interference in crystals composed of regularly spaced atoms.

e) *Schrödinger's wavefunction.* Schrödinger's (Figure 2.5) wave equation describes any object ("system") by a "wavefunction"—an abstract mathematical construct that embodies all knowledge available about the past, present and future of the object at all locations (spatial positions). The wavefunction ascribes wavelike fuzziness to the position of any object and tells us how its spatial contour and extent (wavepacket) change and propagate through time. In particular, a wavefunction may describe diffraction: the splitting of the spatial contour of the object in two or more little-overlapping components when the object encounters obstacles, as experienced by Henry at the doors, and the subsequent recombination of these components.

A wavefunction should describe all the internal and external constituents of the object that evolve independently of each other, known as *degrees-of-freedom.* For an object composed of many atoms that are free to move individually, the number of these degrees of freedom is staggering: it is proportional to the number of atoms. However, if the atoms interact and therefore have correlations with each other, the relevant degrees of freedom may be much fewer: they differ from those of free atoms and are known as the internal *normal modes* of the object. Thus, when macroscopic numbers of atoms vibrate in a crystal, they do so as collective quantized waves known as phonons, with only a few degrees of freedom or normal modes characterized by the frequency and direction of vibration (Figure 2.6).

Fig. 2.5 Schrödinger.

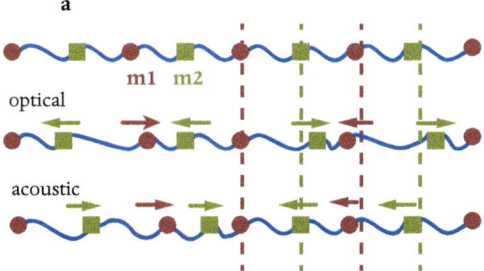

Fig. 2.6 Quantized phonon waves in a crystal. The crystal consists of chains of periodically located atom pairs; e.g., positively charged sodium atoms (circles) adjacent to negatively charged chlorine atoms (squares). The electric pushing and pulling between the atoms and their mass ratio ($m1/m2$) determine the speed and frequency of the compression and stretching of the atomic distances that can be viewed as sound-wave propagation. Such waves may carry energy quanta called phonons.

In Henry Bar's adventure, however, controlled splitting affects only the external degrees of freedom that determine the spatial contour of the object (himself), because these degrees of freedom are totally decoupled from (independent of) the microscopic internal ones.

This has in fact happened in recent experiments by M. Arndt and A. Zeilinger in Vienna, where splitting has been achieved for large molecules, but not quite as large as people (Figure 2.7). The wavelike fuzziness predicted by QM diminishes with the mass of the object, which would make this fuzziness exceedingly small for a person. In addition, the environment obliterates such QM effects and is the more devastating the larger the object (Chapter 8). But prior to analyzing such subtle issues, we ask: What is the principle behind the splitting? The answer is: the superposition principle.

f) *Born's superposition principle is born.* The fundamental maxim of QM introduced by M. Born in 1927 is the superposition principle, whereby any wavefunction can be decomposed into so-called "own functions" or eigenfunctions (in a German–English mix that is the adopted parlance of quantum mechanics).

Each such eigenfunction—an eigenstate—characterizes a measurable value—an eigenvalue—of a physical observable. In the previous adventure of Henry Bar, the eigenvalue corresponded to the door he passed through, while in the case of an electron scattered by few atoms, the atom it was scattered off. If the object is in an eigenstate, the observable (the eigenvalue) remains unchanged as time goes on.

The superposition principle tells us that any combination of these eigenstates, where each eigenstate has an arbitrary permissible coefficient or probability amplitude, is a legitimate quantum state (wavefunction) that can be realized,

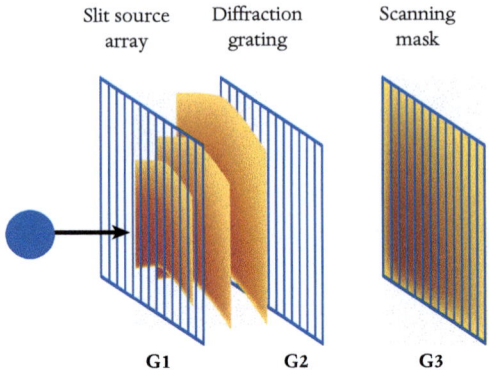

Slit source   Diffraction   Scanning
array   grating   mask

G1   G2   G3

Fig. 2.7 The Arndt–Zeilinger experimental demonstration that large molecules, whose masses range from hundreds to nearly 2,000 atomic mass units, can be split and recombined as quantum waves (provided their translational motion is decoupled from that of internal modes, as in the diffraction experiment in Figure 2.2). The molecules are collimated by array G1, then diffracted by grating G2 and finally recorded at mask G3.

at least in principle. The permissible probability amplitudes in QM are such that the sum of their "weights"—their absolute values squared—must be unity. This requirement, known as *normalization*, reflects the "oneness" of the quantum state. There is no counterpart to this strange concept of a *normalized* quantum superposition in "classical" (electromagnetic or acoustic) wave theory, which also allows for wave superpositions but does not require normalization.

Yet this concept of normalized quantum superpositions has been confirmed by innumerable experiments, some of which we have mentioned previously: Taylor's pioneering experiment, which showed that a single quantum of light (a photon) can still act as a quantum superposition; and Davisson and Germer's experiment, which was the first to show that an electron is scattered off several atoms in a crystal in a correlated fashion, each atom giving rise to a wave with a diminished probability amplitude conforming to the "oneness" of the electron.

In modern experiments, such as those by W. Ketterle (MIT) or M. Arndt (Vienna), a much larger and heavier wavelike object consisting of hundreds or thousands of atoms is split and recombined by an analog of a crystal (a spatially periodic grating) formed by light. These effects will be discussed in more detail in Chapter 3.

g) *Matrix mechanics is born.* A year before Schrödinger's wave mechanics was formulated, N. Bohr's former doctoral student, W. Heisenberg (Germany) (Figure 2.8), had proposed (in 1925) a revolution in physics: his quantum *operator* theory. This theory grew out of his calculation of the transition of an electron

Fig. 2.8 Heisenberg. (Credit: Bundesarchiv, Bild 183-R57262/ Unknown/CC-BY-SA 3.0.)

between two energy states (orbits) in an atom upon absorbing or emitting a light quantum. Heisenberg concluded that in such a transition the changes in electron position and momentum must be described by "operators"—mathematical constructs of a new kind (Chapter 5). The striking consequence of this theory was that, unlike their classical counterparts, the order in which the operators of position and momentum act on the quantum state of the electron can make a drastic difference. This property of operators, called "non-commutativity", will be explained in Chapter 5.

The developments surveyed above, which mainly occurred between 1923 and 1927, constituted a veritable revolution or paradigm change in physics: *the quantum mechanical (QM) revolution*. This book is dedicated to its implications.

2.3 WHY DO SCIENTIFIC PARADIGMS CHANGE?

The dramatic development of QM surveyed above started almost imperceptibly in 1900 from an obscure surmise concerning the need to quantize radiation in a cavity (so as to resolve the mystery of blackbody radiation), and reached its climax in 1927 when the universality of QM was recognized. By a curious and inexplicable coincidence, this scientific revolution was paralleled by another: the advent of relativity, between 1905, when Einstein published his special theory of relativity, and 1919, when Eddington confirmed the validity of general relativity (GR) by his observation of the perihelion of Mercury.

These two scientific revolutions are of similar magnitude. Just as GR has drastically changed our notions of space and time, the QM revolution has completely transformed our world view from its previous "classical" to its current "quantal" version, which is now ninety years old, with no end in sight. Yet the change from the classical to the quantal view is perhaps even more radical, because it transgresses physics and reshapes our basic logic, the meaning of reality or "being" (ontology) and its perception (epistemology), as we will show in subsequent chapters.

Significantly, physicists are still unable to unify or bridge the GR and QM world views, despite relentless efforts which over the past few decades have focused on quantum effects in black holes. Perhaps such "grand unification", should it occur, would bring about a radical revision of the basic QM concepts discussed in this book.

Yet the possibility of such momentous changes in our thinking is puzzling. Is physics not supposed to be (at least since the days of Galileo and Newton) a rigorous, self-critical discipline that strictly adheres to experimental facts and interprets them mathematically? If so, how could this meticulously constructed edifice (based on Newtonian mechanics and Maxwellian field theory) have collapsed as a result of the discovery of a few facts which could not be explained by the ruling theory? And what does the possibility of such an upheaval tell us about our chances to perceive the truth of the world through physics? Perhaps another conceptual upheaval is around the corner.

To cope with these disturbing questions, it is worth our while to revisit the views and facts on the nature of physics and its credibility. Physics is based on intertwined mathematical theory and experimental method. When theory and experiment agree, a natural law is established. The phenomenal success of physics is justly measured by its ability to explain and even predict an astounding scope of experimental facts. But does this success prove that physical laws and concepts hold an undisputed and unique truth?

Such a conclusion is not supported by the history of the development of physics from its origins, which emerged from *mystical, mythical or symbolic views of the world*. Plato's *Timaeus*, the first mathematical "theory of everything", is enshrouded in the mystery of the world creation out of the five perfect geometrical bodies by a creating god. The abstraction and beauty of Plato's theory fascinated modern physicists such as Heisenberg. But was it rationally compelling or yet another myth of creation? The basis of Pythagorean mysticism, before Plato, was that musical and mathematical harmony rule the motion of heavenly bodies. This mysticism was so influential in antiquity that Boethius, the last (sixth century AD) Roman philosopher, found consolation in the "music of the heavens" while awaiting his execution. A millennium later, the laws of Kepler, a Pythagorean, and Newton, a Kabbalist, arose from their quest for a mathematical manifestation of divine wisdom in the universe. Oddly enough, some of the Fathers of quantum mechanics were adepts of mystical schools—particularly Schrödinger, a self-proclaimed Neo-Platonist. This survey suggests that the mathematical language of physics may have roots other than logical necessity. All we can say is that this cultural tradition is vindicated by its undeniable success. Yet the belief in its universal validity is so great that scientists have adopted mathematical forms of communication in their search for extraterrestrial (ET) intelligence. Must such intelligence know of mathematics and its role in explaining the world?

Be it as it may, one cannot deny that the truths of physics are far from eternal: physics has always progressed through paradigm shifts. According to T. Kuhn, each such shift reflects the collapse of the existing consensus within the scientific community out of sociological reasons. K. Popper held the opposite view that such paradigm shifts are inevitable outcomes of the innate logical development of the discipline. In either case, paradigm shifts undermine the strive of physics since ancient times to formulate immutable "super-principles". It is fair to conclude that all super-principles proposed thus far are contestable and have exceptions.

Finally, while physics must incorporate a cohesive explanation to all known facts, this by no means constrains it to a unique theory. This is the case of theories that presume to correct quantum mechanics in currently inaccessible scenarios, such as the one where quantum objects have huge mass or are located in spatial regions where gravity varies abruptly. How then are we to choose among alternative, possibly conflicting, theories if such theories are equally consistent with the experimental observations existing to date? Criteria for such choice that appeal to physicists are beauty, elegance and simplicity, perhaps because they have been nurtured by a tradition that extends as far back as Pythagoras and Plato. Specifically, QM has so far prevailed over competing theories because they all lack these aesthetic qualities. In the following chapters we shall review such alternative theories and point to their aesthetic shortcomings.

We may conclude from the foregoing retrospective that the occurrence of a revolution such as the emergence of QM eludes straightforward explanation. We are in the dark as to when or whether the next one will occur. For many years, the theory of supersymmetry, which purports to unify all possible forces in physics in a common framework based on their symmetry, had been an almost certain bet on "the next big thing" in physics, precisely because it is aesthetically attractive. However, it has recently lost its appeal because of its inability to deliver experimental predictions or contribute to the tantalizing goal of finding a common description for QM and GR.

The QM revolution is still going on, primarily in the realm of novel technologies which it is engendering at present (discussed in Part 3 of this book). These technologies, more than philosophical debates, confront us with deep conceptual issues associated with QM that we shall dwell on. How can we treat many-body complex systems with QM methods? Can we extend QM towards biology, as some researchers contend?

It is conceivable that a radically new understanding of reality will arise from the scrutiny of the issues already mentioned. On the other hand, it is just as possible that the scientific culture of the day, the abundance of erudition and genius among its leaders and the tractable nature of the problems they faced, were unique to the first decades of the twentieth century and may not be repeated at any time soon.

2.4 APPENDIX: SUPERPOSITION, WAVEFUNCTIONS, VECTORS AND MATRICES

In Chapter 1's appendix we introduced the notations of constants and variables, indices and summation. These will enable us to mathematically describe Henry's feats of superposition.

The principle of superposition underlying quantum mechanics can be mathematically formulated as follows. If ψ_1 and ψ_2 are wavefunctions (states) that describe two possible values of a single property of a quantum system—e.g. two positions—then the wavefunction ψ, which is given by $\psi = a_1\psi_1 + a_2\psi_2$, describes a *superposition* of these values. Here, a_1 and a_2 are two variable (complex) numbers that describe the relative contributions of the two superposed states to the overall state or wavefunction. In order to fulfill the constraint that there is but a single object—i.e. it cannot be cloned, duplicated, split or diminished—*normalization* of these numbers is required. Namely, these numbers, known as *probability amplitudes*, can acquire any values as long as the following normalization condition is obeyed:

$$|a_1|^2 + |a_2|^2 = 1$$

where the vertical bars with superscripted 2 denote the squares of their absolute values (i.e. their values irrespective of whether these numbers are positive or negative, real or complex).

This principle was reformulated in a completely different but equivalent language by Born and Jordan, based on Heisenberg's operator theory (discussed previously). The language they adopted to describe quantumness was that of matrices (operators) acting on vectors. Henceforth we shall adopt this language, which has become the predominant formalism of quantum mechanics, and bodes well with the quantum matrix.

Vectors describe states of observables. A *vector* is a mathematical construct that aggregates a discrete number of variables: e.g. a 3-entry vector is denoted by $\mathbf{v} = (a\ b\ c)$, where $\mathbf{v}$ is the vector and a, b, c are complex numbers. As opposed to *lists*, where the order of the numbers does not matter, the position of each entry in the vector is important. Thus, $\mathbf{v} = (a\ b\ c)$ is not equal to $\mathbf{u} = (b\ a\ c)$. Furthermore, vectors can be written either horizontally as a single row, $\mathbf{v} = (a\ b\ c)$, or vertically as a single column, $\mathbf{v} = \begin{pmatrix} a \\ b \\ c \end{pmatrix}$.

Vectors can have an arbitrary number of entries, ranging from one, which means the vector is simply a number, to infinity. Here are some examples of what vectors look like: $\mathbf{v} = (1\ 0)$ is a vector that contains two entries, where the first has the value of 1 and the second the value of 0. $\mathbf{v} = (a_1\ a_2\ \dots\ a_n)$ is a vector with n entries, where the first has the value of a_1 and the nth has a value of a_n.

If vectors represent states of real physical systems, they must be able to describe change in those systems. The state itself can change, but the new state is still represented by a vector (as are all states). How can one describe such a change from one vector to another? The mathematical construct for this is the *matrix*, representing a physical operation—an *operator*. Whereas the vector is a single row or column, the matrix is composed of multiple columns or rows. For example, $\mathbf{M} = \begin{pmatrix} a & b \\ c & d \end{pmatrix}$ is a 2x2 matrix; i.e. it has two columns and two rows. The first column is composed of $\begin{pmatrix} a \\ c \end{pmatrix}$, whereas the second column is composed of $\begin{pmatrix} b \\ d \end{pmatrix}$. In this book we shall mostly use simple 2x2 matrices acting on 2-entry vectors.

In order to change one vector to another, we introduce matrix-vector multiplication. For $\mathbf{M} = \begin{pmatrix} a & b \\ c & d \end{pmatrix}$ and $\mathbf{v} = \begin{pmatrix} x \\ y \end{pmatrix}$, the multiplication proceeds by going through the *rows* of the matrix and summing the multiplication of matrix and vector elements, thus:

$$\mathbf{M} \times \mathbf{v} = \begin{pmatrix} a & b \\ c & d \end{pmatrix} \times \begin{pmatrix} x \\ y \end{pmatrix} = \begin{pmatrix} a \times x + b \times y \\ c \times x + d \times y \end{pmatrix}$$

As can be seen, the result is indeed a new vector:

$$\mathbf{u} = \begin{pmatrix} a \times x + b \times y \\ c \times x + d \times y \end{pmatrix}$$

Some remarks are in order. First, the number of elements in the vector $\mathbf{v}$ must be equal to the number of *columns* in the matrix, otherwise there will be some missing multiplications. Second, the new vector has a number of entries equal to the number of rows of the matrix. Thus, if we wish to transform a vector to another vector of the *same size*—i.e. with the same number of entries—the matrix that performs such an operation has to be square, i.e. number of columns equal to number of rows. Third, here we have used the vertical notation of the vector, but we can also have vector-matrix multiplication where the vector is on the left side, e.g.:

$$\mathbf{v} \times \mathbf{M} = (x \ y) \times \begin{pmatrix} a & b \\ c & d \end{pmatrix} = (x \times a + y \times c \quad x \times b + y \times d)$$

If we have two matrices, $\mathbf{A}$ and $\mathbf{B}$, their multiplication yields $\mathbf{C} = \mathbf{A} \times \mathbf{B}$, which is also a matrix: its number of rows is equal to the number of rows in $\mathbf{A}$, and its number of columns is equal to the number of columns in $\mathbf{B}$. The value of each element is the sum over the multiplications of the columns of $\mathbf{A}$ and the rows of $\mathbf{B}$. For example:

$$\begin{pmatrix} a & b \\ c & d \end{pmatrix} \times \begin{pmatrix} x & y \\ z & w \end{pmatrix} = \begin{pmatrix} ax + bz & ay + bw \\ cx + dz & cy + dw \end{pmatrix}$$

2.4.1 Henry's splitting and recombination as matrix operations

We can now apply the foregoing formalism to the manifestation of the superposition principle in Henry's splitting. Let us first consider his quantum-mechanical passing through the revolving door *and* through the sliding door. Each of these states can be represented by a vector $\mathbf{d} = (r \ s)$, where $\mathbf{d}$ stands for doors, and r (revolving) and s (sliding) are numbers denoting probability amplitudes which tell us "how much of Henry passed through each door". Their absolute squared values represent probabilities according to Born's rule, so that $|r|^2 + |s|^2$ must be equal to one (since there is only one Henry). Since Henry passed through *both* doors, his state can be described by $\mathbf{d} = (1/\sqrt{2} \ 1/\sqrt{2})$, i.e. with one half probability, $(1/\sqrt{2})^2$, he passed through each door, but he did so *simultaneously*, unlike a classical object.

In a more compact notation (introduced by P. M. Dirac) the same vector can be written as $|d\rangle = \frac{1}{\sqrt{2}} |r\rangle + \frac{1}{\sqrt{2}} |s\rangle$, where $|\rangle$ is called a ket. The r-ket, $|r\rangle$

represents the state of passing through the revolving door, and the number in front of it is its probability amplitude; similarly for the sliding-door $|s\rangle$ state.

We now wish to describe Henry's entire quantum exit from his office building in mathematical terms.

We describe his state before hitting the Split button, when Henry was still classical and not yet in a superposition, by the vector $\mathbf{h} = \begin{pmatrix} 1 \\ 0 \end{pmatrix}$; i.e. Henry was then in a single eigenstate.

When he hit the Split button, the change of his quantum state was effected by the 2×2 matrix $\mathbf{S} = \begin{pmatrix} 1/\sqrt{2} & -1/\sqrt{2} \\ 1/\sqrt{2} & 1/\sqrt{2} \end{pmatrix} = \frac{1}{\sqrt{2}} \begin{pmatrix} 1 & -1 \\ 1 & 1 \end{pmatrix}$, with all elements equal to $1/\sqrt{2}$, except one which is equal to $-1/\sqrt{2}$. When Henry's state $\mathbf{h}$ is multiplied (acted upon) by this matrix:

$$\mathbf{S} \times \mathbf{h} = \frac{1}{\sqrt{2}} \begin{pmatrix} 1 & -1 \\ 1 & 1 \end{pmatrix} \times \begin{pmatrix} 1 \\ 0 \end{pmatrix} = \begin{pmatrix} 1/\sqrt{2} \\ 1/\sqrt{2} \end{pmatrix} = \mathbf{d}$$

Henry's state has thus changed from a single state to a quantum superposition of the revolving-door and sliding-door eigenstates where he is in both states at the same time, with equal probability amplitudes.

When he hit the Recombine button his quantum state change was effected by the matrix $\mathbf{R} = \frac{1}{\sqrt{2}} \begin{pmatrix} 1 & 1 \\ -1 & 1 \end{pmatrix}$, where the position of the minus sign is different from that in matrix $\mathbf{S}$. When this matrix multiplies state $\mathbf{d}$, it yields:

$$\mathbf{R} \times \mathbf{d} = \frac{1}{\sqrt{2}} \begin{pmatrix} 1 & 1 \\ -1 & 1 \end{pmatrix} \times \begin{pmatrix} 1/\sqrt{2} \\ 1/\sqrt{2} \end{pmatrix} = \begin{pmatrix} \frac{1}{\sqrt{2}} \times \frac{1}{\sqrt{2}} + \frac{1}{\sqrt{2}} \times \frac{1}{\sqrt{2}} \\ -\frac{1}{\sqrt{2}} \times \frac{1}{\sqrt{2}} + \frac{1}{\sqrt{2}} \times \frac{1}{\sqrt{2}} \end{pmatrix}$$

$$= \begin{pmatrix} 1 \\ 0 \end{pmatrix} = \mathbf{h}$$

thereby restoring Henry to his old, classical self.

It is intriguing to note the following relationship between $\mathbf{R}$ and $\mathbf{S}$:

$$\mathbf{S} \times \mathbf{R} = \begin{pmatrix} 1 & -1 \\ 1 & 1 \end{pmatrix} \times \begin{pmatrix} 1 & 1 \\ -1 & 1 \end{pmatrix} = \begin{pmatrix} 1 & 0 \\ 0 & 1 \end{pmatrix} = \mathbf{I}$$

Here $\mathbf{I}$ is the identity matrix, meaning that when this is operated on a state, nothing changes. This implies that the Split and Recombine buttons/operators do the opposite, and if operated one after the other, cancel each other out.

We will describe Henry's travel through town in more detail in Chapter 3.

The Phases of Henry Bar

What is Quantum Interference?

3.1 HENRY INTERFERES

In Chapter 2 we left Henry in the midst of his search for a means of evading Eve, who stalks him and tries to watch his every move. The Split function on his quantum suit, allowing him to be in several places at the same time—namely, in a non-localized quantum superposition—enables him to perform splendid evasion maneuvers, but it is not enough. Henry wishes to have better control over his actual location so that he may employ his quantum powers to select a route along which Eve cannot see him. For this purpose he has added the phase dial to the quantum suit. This gadget allows an "analog" control over an important quantum variable, called the phase of the superposition state.

In order to understand the importance and usefulness of the phase, we must recall Henry's view of himself as a wave. In the previous adventure, Henry's superposition state was likened to a wave: One cannot pinpoint the position of a wave, which is "spread out", yet it is a single entity, like the superposed Henry. Just as waves in a pond vary between crests and troughs, so do Henry's non-localized states. The height of the wave is its amplitude, and the variation of this amplitude is the phase. Readers familiar with a sine wave know that its zero amplitude ("node") corresponds to a zero phase, its crest (maximal positive amplitude) to a phase of 90°, and its trough (maximal negative amplitude) to a phase of 270°. It is customary to denote the 90° phase as positive and the 270° phase as negative, as they differ by 180° (a full period being 360°).

The phases and amplitudes specify the probability amplitudes of the eigenstates, and thus provide a full description of a quantum superposition. For example, a quantum Henry in a superposition whose phases are all positive is in a different quantum state than almost the same Henry except that one of his superposed states has a negative phase. To change one of the superposed states into another, Henry must turn the phase dial. In order to visually represent this difference we have depicted Henry with a negative phase as the "negative" of the image of his counterparts with a positive phase.

What is the significance of the phase sign; i.e., whether the wave is at its crest or trough? The effect of the phase sign becomes evident when two waves meet and interfere. Just as depicted in Henry's adventure, and as can easily be seen in any pond or bathtub, if their two crests or two troughs coincide, because their phase signs are the same, they reinforce each other, resulting in a wave whose maximum amplitude is the sum of the two waves' amplitudes. This effect is called "constructive interference", as the two waves add up to a larger wave. On the other hand, when a crest of one wave meets the trough of another, because their phase signs are opposite, the result is a flat and "quiet" water surface, as if the water from the crest has filled up the trough. This is called "destructive interference", as the two waves extinguish each other.

Let us apply this wave analogy to Henry's quantum superposition. His superposed states, when combined at the same place, can produce different results, depending on their phases. In Chapter 3 we only depicted the effect of "constructive interference", where both Henry's states—the one going through the sliding door and the one going through the revolving door—had the same (positive) phase sign. Hence, when the different quantum Henrys combined, they "added up" to the full classical Henry who reached his motorcycle. Now the more complete picture unfolds. When Henry pressed the Recombine button, each of the two Henrys moved along two paths. On one path, leading towards the motorcycle, both had the same phase and thus constructively interfered, resulting in a full classical Henry. On the other path, the two Henrys had *opposite* phase signs and thus destructively interfered and "vanished" by cancelling each other out. The result of these two processes was a full classical Henry at one position and a "cancelled" non-existent Henry at the other.

In his latest adventures, Henry's states have changed; one of the superposed states has changed its phase sign from positive to negative. Namely, Henry has changed his wave's crest to a trough by turning the phase dial by 180°. Thus, this dial enables Henry to actively determine which classical Henry will emerge— the one going to the left or the one going to the right. By the 180-degree phase change, Henry has caused the one going to the right to constructively interfere and appear as a classical Henry, whereas the one going to the left resulted in a Henry without amplitude—a null or non-existent Henry whose waves have cancelled out.

This effect is rather odd, since suddenly two of Henry's states have *disappeared*—vanished completely. Such interference of single-particle quantum states is one of the dramatic effects of quantum physics. While interference has been known for centuries for other (classical) types of waves—e.g. water, and acoustic and electromagnetic waves—their interference involves at least two distinct waves. In quantum physics, a *single system interferes with itself.* Henry's interference is not with another person, but rather with other quantum states of himself. The two waves are two quantum *alternatives* of the same system. This can happen only due to the superposition principle: a single quantum system can be in several different states at the same time.

The wave–particle duality discussed in Chapter 2 is made fully manifest now. Henry, who is a single system akin to a particle, behaves like a wave whose main characteristic is interference. Whereas classical acoustic waves interfere because the density of the many particles that make up the medium—e.g. water or air molecules—grows or diminishes at a given location, quantum waves may describe a *single* particle, and yet they interfere in the same way; i.e., they obey the same mathematical description (see Section 3.4, Appendix).

Two mysteries still remain unresolved. The first is the meaning of *absolute phase*, as opposed to relative phase. When two waves meet, their *relative phase* determines whether they interfere constructively or destructively. However, if both waves change their phases in the same way—i.e., their *absolute phase*s change but their relative phase does not—what does this change entail? Apparently, the absolute phase does not matter; i.e., it does not influence any observable property of the quantum system and can thus be arbitrarily set.

Another mystery is the meaning of "paths" as opposed to "states" in our description of interference. This meaning has to do with the dynamical and non-localized nature of waves. Waves that meet do not stop, but rather continue propagating after they pass one another. Their interference, whether constructive or destructive, is either a dynamic change if a propagating wave is in question, or happens at specific locations in space if standing waves are considered. When looking at two waves in a pond, there are specific *locations* at which destructive interference occurs and the water is still. At other locations there is constructive interference. Waves do not change their overall energy on account of interference; they only redistribute their energy in space and time.

The same is true for quantum waves. As discussed previously, Henry is a single entity and cannot duplicate himself or disappear completely due to quantum interference. Hence, when parts of the alternative quantum versions of Henry experience destructive interference, other parts must compensate for it to maintain a whole Henry intact. Thus, Henry's destructive interference is localized in space and time: Henry "disappears" at one place and appears more fully at another. This explains his maneuvers aimed at evading Eve, where two alternative quantum Henrys destructively interfere and become non-existent, only to constructively interfere on the other path where Henry reappears in his full classical glory.

3.2 INTERFERENCE IN QUANTUM MECHANICS

a) *Born's superposition principle and quantum interference*. According to M. Born's superposition principle of 1927, if in a two-slit setup, two wavefunctions describe an electron that emanates from the left or the right slit, then both the sum (constructive interference) and the difference (destructive interference) of the two wavefunctions describe possible physical situations. In fact, they represent mutually shifted patterns on the recording screen behind the two-slit plate: the bright spots (probability maxima) of the waves-sum coincide with the dark spots (probability nulls) of the waves-difference, and vice versa (Figure 2.2).

Interference explains the quantization of energy levels (Chapter 1). If a running wave representing a quantum object (an electron or another particle) is launched inside a confining potential—e.g. a box—it will be reflected off its walls or generally the potential barriers surrounding it. The reflected and the launched waves will interfere, resulting in a standing wave, provided the wave energy has the appropriate value; i.e. it coincides with one of the energy levels (eigenvalues) determined by the confining potential. These standing waves thus have discrete "quantized" energies in Schrödinger's QM. Approximately, but not exactly, they are similar to the standing waves which were at the heart of Bohr's "old" quantum theory—e.g. that of the hydrogen atom energy levels (Chapter 1). Thus, interference is truly central in QM.

Yet the main issue concerning quantum interference is its meaning. What does it mean for a wavefunction to interfere in a statistical ensemble? If this interference implies zero probability of an electron to be present at certain locations on the detection screen, corresponding to the nodes of the pattern, then we can predict with certainty that no electron will hit those spots on the screen. However, all other spots correspond to non-zero probabilities and thus to unpredictable events: a given electron may or may not hit these spots. Yet, crucially, if we apply the quantum–classical correspondence principle to the sub-ensemble of electrons emanating from each slit, we expect them to coincide with a statistical distribution of classical trajectories traversing that slit, but the *phase between the interfering quantum sub-ensembles has no classical counterpart.* In fact, E. Wigner (Hungary, later US) showed in the late 1920s that an attempt to treat such an interfering ensemble as a statistical distribution of classical trajectories must allow for the bizarre notion of negative probability as a signature of its quantumness (Figure 3.1). Overall, the mystery of quantum–classical incompatibility has remained unresolved to this day (Chapter 5).

b) *Ehrenfest's correspondence principle: from quantumness to classicality.* The other maxim that underlies QM is the correspondence principle inferred by the Austrian–Dutch physicist P. Ehrenfest based on M. Born's work. This is a recipe as to how one obtains classical (Newtonian) behavior from a quantum (Schrödinger's wavefunction or Heisenberg's operator) description. Here too, the notion of a statistical ensemble provides a rather obvious recipe: classical behavior of any observable—e.g., position, momentum or energy—corresponds to its average (mean value) over the ensemble. This means that we weigh the quantum observable by its probability to have a particular value and average over all such values. Since probability in QM according to Born's rule is the square of the probability amplitude that characterizes the wavefunction, this recipe implies that the more diffuse or spread out the wavefunction in space, the smaller the probability that its position has a certain value. Thus, a wave uniformly distributed in space has an undefined mean position. On the contrary, a wave

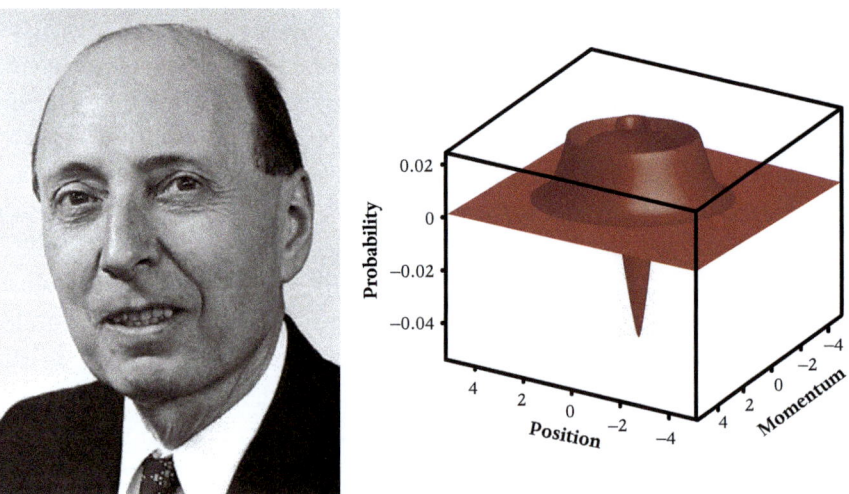

Fig. 3.1 (Left) E. Wigner. (Right) Wigner's quantum-mechanical analog of a statistical distribution function in position and momentum may have negative parts, as opposed to a classical statistical distribution that measures their (necessarily positive) probability.

localized around a certain location has this location as its mean position. Since (Chapter 1) the (de Broglie) wavelength of quantum waves becomes exceedingly small when particles much more massive than an electron are considered, such waves are strongly localized around their mean positions, rendering the classical limit highly accurate.

c) *The debate on quantum interference.* As soon as Schrödinger's equation for the wavefunction and Heisenberg's equation for operators were recognized in the late 1920s to be equivalent and hold universally, a stormy debate broke out over the meaning of the "new" quantum mechanics (QM), as it was termed. In a series of congresses held in Solvay, Belgium, from the mid-1920s to the early 1930s, a controversy raged between the "realists"—mainly Einstein and de Broglie— and adherents of the Copenhagen interpretation on the meaning of quantum interference.

The realists insisted that a proper theory must describe what "really" happens to the object one examines. For example, when an electron hits a plate with two distant slits, theory must be able to tell which slit the electron goes through. They did not accept the inability of Schrödinger's wavefunction to reveal this slit unless a detector is placed at the slit, but then according to QM the electron is no longer described by the same wavefunction, as will be shown in Chapter 4.

De Broglie went so far as boldly suggest that QM be amended to make it conform to his "realism". In his theory, a "pilot wave" that carries no energy guides the electron through the slit that it actually traverses, whereas the Schrödinger wavefunction determines the distribution of energy and matter of the electron (or any other physical influence the electron may exert) in space and time. This theory of a "covert" pilot wave was the precursor to the subsequent theory of "hidden variables" that D. Bohm (US, later Israel and UK) developed in the 1940s and 1950s. These theories reflected the view that QM does not suffice for a complete description of the world—that it must be complemented by a deeper level of description in terms of mysterious variables that elude detection. This view has been repeatedly refuted by experiment, and there is no way to reconcile it with the general framework of QM.

The followers of the Copenhagen interpretation developed by N. Bohr with much input from W. Heisenberg and M. Born, which soon became the "official" interpretation of QM, viewed the wavefunction not as a description of an event, such as the traversal of the two-slit plate by an individual electron, but as a *statistical* description of many events (a "statistical ensemble"). Thus, knowing the wavefunction does not suffice for answering the question as to which slit was traversed by a particular electron. Bohr maintained that the very question cannot be asked, because it pertains to the "unknowable" in QM. His insistence that questions concerning the *unknowable* in QM must not be asked provoked the humorous reaction of Wigner: "What would happen to me if I did?"

The distinction between the knowable and the unknowable shifted the focus of QM interpretation to the limitations of our knowledge. It became clear that on the one hand the wavefunction cannot fully specify *all possible observables*—e.g., the position of each electron with respect to the slits. as well as its direction of propagation (Chapter 5). On the other hand, the wavefunction cannot be fully known from a single event, such as the traversal of the two-slit plate by a single electron that leaves its record on the screen behind the plate.

In 1929, J. Von Neumann (Hungary, later Germany, then the US) connected the QM description in terms of a wavefunction with the notion of information, which had already existed in the nineteenth-century statistical theory of classical (Newtonian) particles—statistical mechanics. In nineteenth-century statistical mechanics, the lack of information (or the ignorance) about the positions and velocities of the multitude of microscopic particles in the ensemble was identified with the entropy of the state of the ensemble by Gibbs (US) and Boltzmann (Austro-Hungary). By the same token, Von Neumann associated entropy in QM with the lack of information concerning the quantum state of the object, so that if the state is known we have zero entropy. However, a known state signifies full information only in the QM sense, but still does not allow us to answer questions on the unknowable in Bohr's sense, such as the complete route of a single electron in the two-slit setup.

d) *The spin as an interfering two-state system.* The debate over the interpretation of QM in the late 1920s and early 1930s revealed the need for very simple objects whose quantum features could be fully and lucidly analyzed. In the first years of QM, discussions revolved almost exclusively around continuous wavefunctions—i.e. wavefunctions whose probability amplitude varies smoothly in space—so that each spatial point in the wavefunction must be specified for us to fully comprehend their interference properties. Such wealth of information corresponding to an infinite number of degrees of freedom is hard to handle, both experimentally and in terms of analysis. It was therefore highly beneficial for understanding the basics of QM that during those years a remarkably simple type of quantum systems, with only two degrees of freedom or two eigenstates, was discovered: the *spin*, an internal rotation of a quantum particle, first predicted by G. E. Uhlenbeck and S. A. Goudsmit (Holland) in classical form, and reformulated quantum-mechanically by W. Pauli (Switzerland).

The existence of the spin was experimentally demonstrated by O. Stern and W. Gerlach (Germany) for an electron (Figure 3.2). The Stern–Gerlach

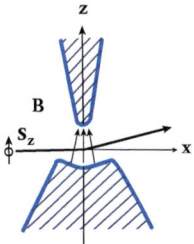

Fig. 3.2 The Stern–Gerlach setup for spin detection in a beam of silver atoms. The variation (gradient) of the magnetic field B aligns the spin component of the atom with z and causes the deflection of its momentum from the x axis towards the z axis.

experiment consisted in the propagation of a beam of silver atoms through a magnetic field that grows (has a gradient) in the direction perpendicular to the beam propagation. The magnetic field exerts a force on the atomic spin, causing atoms with spins pointing up and down to deviate in opposite directions. The spin magnitude was deduced from this deviation.

The peculiar quantum-mechanical property of the spin is that it turns the particle into a tiny magnet when placed in a magnetic field. However, unlike a classical rotator it has only two energy values (eigenvalues) or levels: the higher level is for a spin pointing up (along the field), and the lower for a spin pointing down, thus forming a two-eigenstate system. The reason that an electron has only two energy eigenstates for the spin is again quantization of the action in units of $\hbar$: there can be no intermediate orientation of the spin between "up" and "down", since such orientations would differ by less than $\hbar$ in the action required to change the spin direction.

Atomic nuclei have spins as well, as was demonstrated by an effect termed *nuclear magnetic resonance* (NMR), discovered in the 1940s by I. Rabi (USA), who employed electromagnetic pulses to make the spin rotate (precess) at a fixed angle about the axis of the static magnetic field. This precession occurs only when the electromagnetic pulse frequency matches (is resonant with) the splitting between the energies of the spin states that point up and down along the magnetic-field axis. Each pulse exerts a kicking force in the direction perpendicular to the static magnetic field, so that it causes a torque on the spin, as if it were a push of a spinning top.

The precessing spin is described in QM by a superposition of the up-pointing and down-pointing spin eigenstates with time-changing (oscillating) probability amplitudes (Figure 3.3, and Section 3.4, Appendix). Born's rule implies that this precession corresponds to an oscillation of the probability of finding the spin point up or down, and thus having a higher- or lower-energy eigenvalue. The energy oscillation of a macroscopic ensemble of spins was measured by Bloch and Rabi (Figure 3.4).

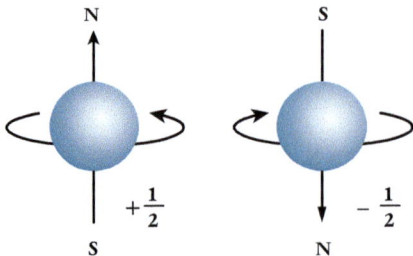

Fig. 3.3 Electron spin: visualization as a spinning top that generates a magnetic field and has a value of either $+\frac{1}{2}$ (spin-up) or $-\frac{1}{2}$ (spin-down).

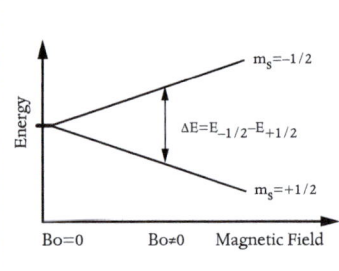

Fig. 3.4 (Left) F. Bloch, the discoverer of spin precession. (Credit: Stanford University/Courtesy Stanford News Service). (Center) I. Rabi, the discoverer of the nuclear magnetic resonance (NMR) effect. (Right) Energy splitting of nuclear spin levels in a magnetic field—the origin of the NMR effect.

Because energy splitting is sensitive to the magnetic field near the measured spins, NMR has become an invaluable tool for studying the material composition at tiny distances from the atoms or molecules whose nuclei carry the spins, as the spins are highly sensitive probes of material properties, with applications in chemistry, physics and medicine.

With the advent of powerful radiation sources at microwave frequencies, known as "masers", in the 1950s, it became possible to manipulate superpositions of two rotational states of ammonia molecules that have the appropriate energies. Since the 1960s, optical lasers have allowed similar operations on atoms with pairs of electron energy levels separated in energy by $\hbar$ times the laser frequency.

These manipulations of the two-state spin system and their analogs in atomic, molecular and photonic systems have become central tools of QM research (discussed in subsequent chapters). In particular, such manipulations help us to clarify the notion of quantum interference between the two states of the system (Section 3.4, Appendix).

e) *From matter waves to quantum information.* Schrödinger's wave mechanics emphasized motional effects of quantum waves. The realization that this description is universal and that the wavefunction carries information rather than a physical influence was introduced by the astoundingly deep work of the Hungarian-born (later American) scientist J. Von Neumann between 1929 and 1932 (Figure 3.5).

This work has laid the foundation of what is currently known as quantum information theory, which draws on the fact that a quantum-superposition state may encode much more information than its classical counterpart. The building block of this theory is a quantum system that can occupy only two (orthogonal) eigenstates, just like the spin described previously, or its analog; e.g., an atom with two internal states of different energy or a quantum of light (a photon) with two orthogonal polarization states. (In Henry's tale the two states are his versions passing through the revolving or sliding door.) If we label the two states 0 and 1, such a system is reminiscent of a logical bit of information in a computer.

Fig. 3.5 J. Von Neumann, the Father of quantum information theory.

Its quantum counterpart is called a "quantum bit" or qubit. Contrary to the classical bit, a qubit (like Henry) can be in *both states*, 0 and 1, at the same time. In subsequent chapters we shall discuss what can be done with such a weird information-carrying system.

f) *Quantum interferometry*. The superposition principle applies to two-slit interference setups, where a quantum wave can propagate along two alternative, indistinguishable paths, each corresponding to a path-localized eigenstate, the two eigenstates being orthogonal to each other. Their phase difference (relative phase) is the crux of the quantum interference. To focus on this aspect, let us consider two-slit interference in which the probability amplitudes of the two superposed eigenstates, corresponding to traversal of the left and right slits, have equal magnitude (amplitude) but possibly different phases.

The superposed states can be visualized as vectors (Section 2.3). Their relative phase θ is determined by the difference between the alternative propagation paths 1 and 2 (Section 1.3). The relative phase that renders the two-state interference destructive or constructive corresponds, respectively, to an odd or an even multiple of π (180°). The two states interfere constructively for $\theta = 0$, so that their probability amplitudes add up, or destructively, so that their probability amplitudes are subtracted, for $\theta = 180°$.

It is easy to interpolate between these cases by varying θ. The endeavor over the years has been to build devices called interferometers which can precisely determine the relative phase that alternates with a period of the de Broglie wavelength λ_{dB}, although this wavelength is extremely tiny for objects much heavier than an electron. Nevertheless, ingenious designs have recently allowed measuring the phase in two-slit interference for molecules that weigh more than 1,000 atoms (by M. Arndt in Vienna, Figure 2.7), and for ultracold atom clouds with similar mass (by W. Ketterle at MIT).

An alternative observation of two-slit interference can occur in *time* rather than in space. Such an effect was predicted by G. Kurizki and A. Ben Reuven (Tel Aviv, 1985) and experimentally demonstrated by P. Grangier, A. Aspect and J. Vigue (Paris, 1985) for two identical atoms. These atoms formed a molecule that was dissociated upon absorbing a photon. As the two atoms receded from each other, the photon was re-emitted by one of the indistinguishable atoms, analogously to two-slit interference with the atoms in the role of "slits". Since the distance between these "slits" grew in time as the atoms receded, the relative phase evolved from 0° to 180°, causing the interference to vary from constructive to destructive.

3.3 THE DEEPER MEANING OF QUANTUM SUPERPOSITIONS

a) Quantum Hamletism: to be and not to be

The quantum superposition principle allows an object to coexist in states of "being" and "non-being" at a given place and (or) time. This possibility has no precedent or analogy in the older "classical" physics.

Such coexistence is the essence of quantum wavelike propagation (diffraction or interference) of an object in two-slit setups, where we cannot claim the object has passed through ("existed in") any particular slit.

In the time domain, such coexistence has already led to spectacular manifestations of the "rebirth" of quantum-superposition states that are presumed "dead"—i.e. non-existent—at certain times, as described in subsequent chapters.

Such phenomena are being pushed to the realm of large, complex objects: superconducting circuits, nanomechanical springs or atom "clouds". Before long, they may be observed for *live objects*, such as viruses or bacteria. Then, *simultaneous* existence and non-existence will have a particularly striking connotation.

Space-domain and time-domain superpositions of "yes" and "no" states are at the heart of the revolutionary quantum logic, which is the basis of extensively developed quantum computers and information transmitters, as explained in subsequent chapters.

Such coexistence of "being" and "nothingness" appears bizarre, not only to our daily experience but also in the realm of philosophical speculation, since it contradicts the basic dichotomy between these two notions postulated in Western philosophy 2,500 years ago by the Greek philosopher Parmenides, who stated: "What is, is, and what is not, is not", so that being cannot become (let alone coexist with) nothingness.

Yet such coexistence appears to resonate with the teachings of the great Buddhist thinker Nagarjuna (second century AD), to whom both being and nothingness are as much void of meaning as their negation or any other combination. We contend that this defiance of common logic or experience by Nagarjuna and his Buddhist followers is truly in the spirit of the quantum superposition principle!

If indeed being and nothingness coexist in a quantum state, does this mean that nothingness is real? The *reality of nothingness* has roots in the Platonic School of philosophy, which inspired mysticism in all monotheistic religions from the second century BC to the Middle Ages. Platonists asserted that the Infinite is Naught, yet paradoxically it is the source of Being. Strikingly, similar ideas were expressed in the second century BC by the Chinese thinker Lao Tze, the founder of Taoism, in his discourse on the Tao, the essence of the universe.

Such ideas about the reality of nothingness have persisted to the modern era; e.g., in Heidegger's philosophy. Lewis Carrol shrewdly alluded to them in his book *Alice Through the Looking Glass*, where Alice tells the White King "I see nobody on the road", to which the King replies "I only wish I had such eyes. To be able to see Nobody!"

Philosophical speculation on these issues may be irrelevant to quantum physics. Rather than indulge in such speculation, we may confine ourselves only to issues that have operational meaning (to be pursued in Chapter 4), namely: what measurement outcomes are allowed by QM in a given setup? Yet even if we shun from philosophical discussion, quantum *logic* that is based on superpositions of "yes" and "no" is here to stay, emerging in the age of quantum information, so that we may be confronted with its implications for our lives. Accordingly, a quantum-age Hamlet would perhaps resolve his famous dilemma as follows.

The quantum Hamlet

To be *and* not to be, that is the answer:
Suffer the slings of outrageous fortune
And yet take arms against a sea of troubles . . .
To die and live, to sleep and wake at once – 'tis our quantum lot.

b) To know and not know

The strange notions of quantum states or wavefunctions may be either dismissed or taken in earnest, depending on our answer to a single question. Does an object (electron, atom and so on) actually *have* a wavefunction? Probably not, we may answer, as a wavefunction is not a property. Its interpretation in terms of *what is "knowable" on the object* shifts the focus from the object towards the potential gatherer of this knowledge or information: the observer. As we shall see in Chapter 4, the observer plays a central role in QM, according to most interpretations. Does this role of the observer make QM more anthropocentric (human-centered) than other theories? Not necessarily. Statistical mechanics and thermodynamics have given rise to the concept of entropy, which expresses ignorance. Physical concepts are abstract rather than directly linked to our experience, despite the reservations of E. Mach and other positivists. The American physicist Bridgman advocated in the 1940s *operationalism*, but included in it "pen

& paper operations"—meaning theoretical concepts—as an essential part of physical research, even though such concepts may not be directly operational.

If we adopt the Copenhagen interpretation of QM, we are confronted with Bohr's view that there are constraints inherent to human comprehension (epistemological) on *what is knowable and thinkable*: for example, where was the particle in a two-slit setup prior to its detection? Clearly, we need not accept the existence of such constraints. However, we cannot disengage ourselves from the conceptual limitations of QM, which indeed cannot answer the this question without destroying its framework altogether. Are we then in a logical impasse as regards the possibility of overcoming the constraints posited by Bohr on QM? Not necessarily! In Section 5.3 we shall explore a possible way out.

3.4 APPENDIX: INTERFERENCE AND QUANTUM WAVES

In this appendix we continue our formulation of quantum states and describe in more detail the concepts of superposition, state amplitudes, phases and interferences. Henry's state in his apartment can be described by a superposition of two distinct classical states: sitting (S) and leaning (L):

$$\psi = a_S \psi_S + a_L \psi_L$$

with the normalization $|a_S|^2 + |a_L|^2 = 1$ that maintains a full (single) Henry. In Chapter 2 the probability amplitudes acquired only equal values, but in a general quantum state they can have *any value* as long as their normalization is maintained. Note that the normalization involves the *absolute valuessquared of the amplitudes* and not their values. Since the sum of the absolute-value squared amplitudes must equal 1, the amplitudes themselves can be negative. By contrast, for *probabilities*, what counts are their values (which must be zero or positive), and the sum of probability values must equal 1. This simple distinction between probability amplitudes and probabilities is one of the cornerstones of quantum physics and the origin of many counterintuitive effects that will be discussed in detail in this book.

In the spirit of Henry's understanding of these phenomena, let us consider, for example, the amplitudes of waves whose mathematical forms are sines and cosines (Figure 3.6).

Waves have three main characteristics: their amplitude (height), phase (*x*-axis positions of the crests and nodes) and wavelength (distance between two crests). Mathematically, a wave can be expressed as:

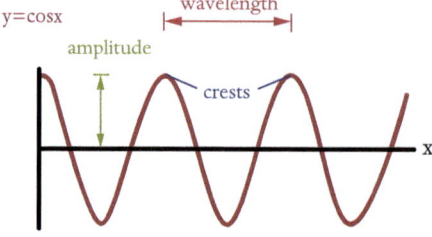

Fig. 3.6 A cosine-wave amplitude (*y*) as a function of *x*.

$$a = \text{amplitude} \times \cos(\text{phase} + x/\text{wavelength})$$

There are several mathematical (trigonometric) relations that are of importance to us. The first is that the only difference between cosine and sine is the phase: $\cos(x) = \sin(x - 90°)$. The second is $(\sin x)^2 + (\cos x)^2 = 1$, and the third is $\cos (x - 180°) = -\cos(x)$.

With these mathematical relations we can describe quantum interference. As discussed in detail in this chapter, interference occurs when two waves meet. Let us consider two unity-amplitude waves (amplitude $= 1$) of the same wavelength λ, which differ only in their phases:

$$a(x, \text{phase}_1) = \cos(\text{phase}_1 + x/\lambda)$$
$$a(x, \text{phase}_2) = \cos(\text{phase}_2 + x/\lambda)$$

What happens when they meet? If they have the same phase—phase$_1$ = phase$_2$ = φ—then constructive interference occurs: $a(x, \varphi) + a(x, \varphi) = 2 \times a(x, \varphi) = 2 \times \cos(\varphi + x/\lambda)$. The sum-wave has exactly the same form, only with *double the amplitude*: its crests are twice higher and its troughs are twice lower than those of each wave.

If the waves have opposite phases, which occurs when Henry turns the phase dial to 180°, so that phase$_1$ = phase$_2$ − 180°, then these waves interfere destructively:

$$a(x, \text{phase}_1) + a(x, \text{phase}_1 - 180°) = a(x, \text{phase}_1) - a(x, \text{phase}_1) = 0$$

where we have used the fact that $\cos(x - 180°) = \cos(x)$, which means that a phase change by 180° is equivalent to a sign-change of the amplitude. When the two waves meet, each crest meets its opposing trough (and vice versa), and they cancel each other out. This destructive interference can only befall part of the entire state: it must be compensated by constructive interference of the other part, so as to preserve the "oneness" of the object, as illustrated subsequently.

From this simple example we can understand why Henry named his new quantum suit contraption the phase dial. The phase controls which type of superposition occurs: constructive, destructive or anything between.

Let us now describe Henry's full state in his apartment (as well as to evade Eve) and the full dynamics that resulted in an interference/disappearing act. For simplicity we shall omit the x-dependence of all the probability amplitudes (in later chapters they will play a crucial role). Henry's initial state can thus be described as:

$$\psi_S = \begin{pmatrix} 1 \\ 0 \end{pmatrix}, \psi_L = \begin{pmatrix} 0 \\ 1 \end{pmatrix}$$

$$a_S = a_L = \frac{1}{\sqrt{2}}$$

$$\psi = a_S \psi_S + a_L \psi_L = \frac{1}{\sqrt{2}} \begin{pmatrix} 1 \\ 1 \end{pmatrix}$$

Henry's leaning quantum version turned the phase dial to 180°. This, as we have seen, resulted in a sign change of the leaning state amplitude only. This action can be formulated by a quantum operator as follows:

$$P(\varphi_L = 180°) = \begin{pmatrix} -1 & 0 \\ 0 & 1 \end{pmatrix}$$

Applying this operator on Henry's state results in:

$$P(\varphi_L = 180°)\psi = \frac{1}{\sqrt{2}} \begin{pmatrix} -1 \\ 1 \end{pmatrix}$$

where now the leaning quantum state of Henry has the opposite sign to that of the standing one.

Next, Henry's quantum versions moved towards each other, whereupon they recombined. The recombination operator, presented in Chapter 2, is denoted by **R**. Let us see in detail what happens to Henry's state as it is recombined:

$$\mathbf{R} \times P(\varphi_L = 180°)\psi = \mathbf{R} \times \frac{1}{\sqrt{2}} \begin{pmatrix} -1 \\ 1 \end{pmatrix} = \frac{1}{\sqrt{2}} \begin{pmatrix} 1 & 1 \\ -1 & 1 \end{pmatrix} \frac{1}{\sqrt{2}} \begin{pmatrix} -1 \\ 1 \end{pmatrix}$$

$$= \frac{1}{2} \begin{pmatrix} 1 \times (-1) + 1 \times 1 \\ (-1) \times (-1) + 1 \times 1 \end{pmatrix} = \frac{1}{2} \begin{pmatrix} 0 \\ 2 \end{pmatrix} = \begin{pmatrix} 0 \\ 1 \end{pmatrix}$$

Thus, the upper row represents destructive interference ($-1 + 1 = 0$), whereas the lower row represents constructive interference ($1 + 1 = 2$). Contrast this with the Chapter 2 calculation, which resulted in the state $\begin{pmatrix} 1 \\ 0 \end{pmatrix}$. The only difference between the two scenarios is the change in the sign of the superposed state. This example shows how a change in phase can drastically change the state by converting destructive to constructive interference, or vice versa.

60

The Collapse of Henry Bar

A week later. Henry, with Johnny's help, solved the equations

Eve has hacked into the station's surveillance system.

I can't trust the solutions I've written on the board to any computer, I'm sure Eve can. tap them!

The station camera is on.

I'll split to get Eve off my track.

Where is he?

What happened ?!

One version of me vanished, I'm glad I'm still whole, though.

CHAPTER 4

What are Quantum Measurements?

4.1 HENRY IS MEASURED

This chapter introduces us to one of the greatest mysteries of quantum mechanics: the notion of quantum measurements. In previous chapters, Henry had split into several quantum versions of himself which interfered as they recombined. However, in our daily lives, or even in the most sophisticated physical laboratory, we never *measure* anything that is located at two places or has two different values at the same time. When we observe and measure any physical observable we find one or another result, but not both at once; it may be a specific location, a specific momentum (the direction and amount of motion) or an energy level. We never measure a superposition of any observable values of a quantum system. A snapshot of a superposed photon always records a single spot on the detection screen, never two, nor half a spot. If so, *do superpositions really exist*, even though we cannot measure them?

To answer this question we need to re-examine what we have learnt thus far about Henry Bar's quantum suit and its principles of operation. When Henry is in a superposition state he is physically in several places, but just like any other quantum system, when Eve *measures* Henry's position by placing a video camera she (as a classical observer) must have a *single result*: either she sees Henry *or* she does not. Since Eve is not a quantum system (yet) she cannot be in a superposition of detecting Henry *and* not detecting him at the same time. Hence, a measurement by a classical observer produces a single result. Furthermore, once Eve has detected Henry on the video camera at a specific location, he *must* be there. Our physical reality is made of these measurements. If something is measured somewhere, then it exists at that location; it cannot be anywhere else. Thus, Eve's measurement of Henry *forces* Henry's position to be classical.

But what is a measurement? Is mere looking at an object deemed its observation? And who or what is an observer? Is the camera an observer, or must there be *someone* to watch the video? If so, will Henry collapse when Eve is not looking at the camera? Such issues are still being debated, and we discuss current standpoints on these issues in Sections 4.2 and 4.3. So, what *do* we know about measurements?

We know what they do to superpositions. Prior to his measurement, Henry's quantum state had been described by a wavefunction that is non-localized and can interfere with itself. However, once Eve detected Henry at a specific location he ceased to be wavelike; on the contrary, his spread-out position *collapsed* to a classical, particle-like localization. This effect of measuring a superposed quantum state is thus called "wavefunction collapse".

Yet something even stranger happened following Eve's first attempt at detecting Henry. One of Henry's quantum versions (lookalikes) passed in view of Eve's camera, while the other did not. Eve *did not* detect Henry at the camera, yet Henry's state nonetheless collapsed to its quantum version that had not passed in front of the camera. Why did Henry collapse if Eve had not detected him? The answer is that *a negative result is still a result*. By Eve's non-detection of Henry's version in front of the camera, she collapsed his state to *not being* in front of the camera. Otherwise, Eve's perception of reality would have contradicted Henry's state. Thus, by detecting the *absence of Henry*, she made him collapse to the version that was not there and made him classical again.

Which state does the wavefunction collapse to? Who or what decides if it is this or that state? These questions have puzzled some of the great minds of our times. There is one clear piece of evidence concerning this conundrum: the "decision" as to which state the wavefunction collapses to is completely, utterly *random*. In fact, some physicists claim that this randomness is *the only* randomness in the world. There is no way to influence or predict the result of a particular collapse event; there is no certainty as to what will happen if you measure a superposed state. Henry comes to this conclusion after his unexpected collapse, as he measures his own superposed self by taking many selfies, and observes that each time, a different version of him disappears. No matter how hard he tries to analyze and predict the pattern of collapse, he finds there is simply none.

There is still a chance for a prediction, albeit uncertain, because each superposed state has a number associated with it: the probability amplitude discussed in Chapter 3. The absolute-value squared of this number determines the *probability* that the wavefunction will collapse to that particular state. Thus, if Henry is split into two equal probability-amplitude quantum versions, there is a 50% chance for each version to appear after a measurement. By contrast, if Henry splits in two quantum versions with unequal probability-amplitudes—say, one whose square is 20% and the other whose square is 80%—then the former version has a lower probability of appearing after a collapse than the latter by a ratio of 1:4. Here, probability means that if Henry repeatedly performs the same operation—say the Split operation followed by a measurement—then the spread of the results after many repetitions will be 20% for the former version and 80% for the latter. Thus, a superposition state is fully described not only by the superposed states, but also by their corresponding probability-amplitudes. We remind the reader that the phases of the probability-amplitudes discussed in Chapters 2 and 3 are also essential for the description.

In this chapter, Henry was defeated by his arch-enemy Eve. His quantum suit did not offer any protection against Eve's measurements. This raises a major philosophical issue concerning the role of the observer in quantum physics: apparently, *an observer is never passive*. The mere possibility of measurement owing to the presence of an observer changes the state of the observed object. Just by trying to observe Henry, Eve has managed to change his state from quantum, here wavelike and spread-out, to classical-like and localized; even though she did not detect him. With notable exceptions discussed in Section 4.2, measurements change the quantum state. Can Henry prevail against such a brutal attack on his quantum powers? Subsequent chapters will tell.

4.2 QUANTUM MECHANICS AS A MEASUREMENT THEORY

M. Born explained in 1927 that QM is probabilistic because it is essentially a measurement theory, in which a quantum state pertains to a large collection (ensemble) of identical objects or copies (e.g., electrons, atoms, photons etc.); hence, the behavior of a single object (copy) is unpredictable within this theory. In this view, adopted by N. Bohr in his Copenhagen interpretation of QM, the quantum state is merely a catalog of possible measurement results, once the observer decides which observable is to be measured. If the measured observable is in a superposition of eigenstates, then the measurement result each time is random and unpredictable. Our chances of guessing the result correctly beforehand are given by the probability of finding the corresponding eigenstate in the superposition state. The notion of unpredictability/randomness was put in focus with the advent of the Stern–Gerlach setup as a platform for quantum measurements. Thus, if the spin of the silver atom in the setup is aligned with the z axis, then it is in an equal superposition of the spin aligned with $+x$ and $-x$

(Chapter 3), so that its detection along $+ x$ will only succeed with a 50% probability, i.e. in half of the trials, without any clue as to which trial will or will not succeed.

However, a key ingredient was still missing in this interpretation: how to describe the act of measurement in QM. The answer was given in 1932 by J. Von Neumann as part of the linkage he created between QM and information theory. His first step was to extend the wavefunction concept to a statistical function of a quantum ensemble that he named the density matrix (which was concurrently introduced by the Soviet physicist L. D. Landau). From the density matrix we can infer the entropy or the information stored in the ensemble, which is a measure of how close it is to a "pure" superposition state (that corresponds to zero entropy, i.e. maximal information), and how different from a statistical "mixture" of its eigenstates that corresponds to a higher entropy (less information), wherein each eigenstate is characterized only by its probability (statistical weight). Such a "mixed" state has no dependence on the phases that are essential in a pure superposition state (Chapter 3).

Von Neumann's crucial step was to identify a measurement with a sharp transition, caused by the measuring apparatus, from a pure state to one of the eigenstates in the superposition, followed by the readout of the observable value that corresponds to this eigenstate (Figure 4.1). He postulated that such a transition, which he termed *wavefunction collapse*, must occur if a single result (observable value) is recorded by the measuring apparatus each time, albeit at random. Yet he stressed that one cannot deduce this collapse from QM rules (the Schrödinger equation), whereby a wavefunction initiated in a pure superposition state will remain in such a state. In information theory parlance, the collapse is an abrupt jump from partial information about the presence of each eigenstate in the super-position to full information about a single eigenstate. In Copenhagen-school

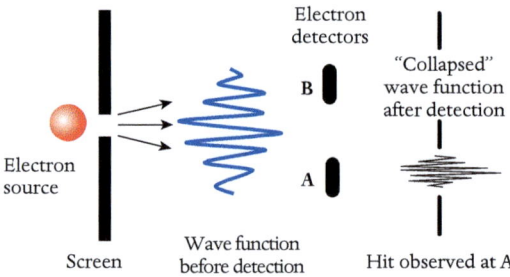

Fig. 4.1 Schematic visualization of Von Neumann's wavefunction collapse (measurement postulate).

jargon, the collapse realizes one of the potentialities of the system and turns it into an actuality.

Von Neumann's measurement postulate whereby collapse cannot be deduced from the Schrödinger equation was a landmark in the conceptual development of QM, but the perplexion it has created is still with us today. Despite the extensive elaboration on the measurement process over the years, we still have not fully reconciled the notion of collapse with QM rules, as will be discussed later in this chapter and in Chapter 5. The main unresolved issue is: where and when does a wavefunction collapse? Von Neumann unequivocally replied, *in the mind of the observer*—the person who reads out the measurement result. His radical position has been vehemently rejected by many (but not all) researchers as being non-physical or even non-scientific. We shall review this debate in Chapter 5, but the question is: are there alternative solutions to the collapse problem?

Von Neumann charted the path that has been followed by those that have been trying to "banish the observer" from QM. This path consists in a quantum-mechanical description of the measuring apparatus that, in the wake of its interaction with the measured system, becomes correlated with it. According to Von Neumann, the states of the system and the apparatus are then *inseparable*: they were termed *entangled states* by Schrödinger in 1935 (see Chapter 7). In order to determine the ensuing state of each of these two correlated or entangled entities, we have to decide what to do with the other. Do we specify the state of the system upon ignoring the apparatus, which amounts to averaging over all possible states of the apparatus (a process known as an unread measurement)? Alternatively, do we select a state of the apparatus, and thereby single out the corresponding state of the system correlated to it (resulting in a projected state and a genuine measurement)? And who decides which of these alternative processes is realized?

A resolution of this conundrum proposed by several researchers over the years, notably by W. Zurek (USA) in the 1990s, has been that no decision or human intervention is called for, as the measuring process is determined by natural causes which he termed *quantum Darwinism*. According to this approach, the apparatus and the system end up in the most *robust states* available to them, the guiding principle being *survival of the fittest states*.

A good example of this approach is the spontaneous emission of a photon by a two-level atom that is initially in its excited (upper) energy state. The detection of the emitted photon signals the transition of the atom to its lower state, which is the robust (stable) state of the two. Yet even if the photon detector has not "clicked", let alone been read out by an observer, the measurement has occurred as soon as the photon has been emitted. However, there is a caveat to reckon with: the emission process is not instantaneous, nor does the photon–atom correlation vanish at once. It lasts for a time that is typically extremely short except for situations wherein the photon and the atom are confined between highly reflective mirrors; i.e., in a cavity. The emitted photon then bounces back and forth between the mirrors, repeatedly crossing the atom and thus reviving their correlation. Then, an observer/controller may intervene in the process by acting on the atom, causing a reversal of the process—an "undoing" of the measurement. Hence, a "natural" measurement that ends up in the "fittest" state is not unavoidable, as the observer may choose to abort it while it is happening. Does this imply that the observer cannot be banished from QM?

A highly publicized approach to the measurement problem in QM has been H. Everett's (1957, USA) *many-world or parallel-universes* interpretation, which we shall describe in Section 4.3. In this approach, the measurement of a superposition state *splits our universe* such that in each of the "new" universes there is only one of the superposed states and hence only one result.

Another approach was advanced by R. Penrose (UK) in the 1990s, whereby gravity may act differently on each component (branch) of a quantum superposition state that is spread out in space: In a sense, gravity is a kind of observer or measuring apparatus of such quantum objects. Very recent works by J. Maldacena (USA) have carried this approach much further. Yet it seems safe to assume that the differences in gravity action on different branches of a typical quantum superposition state created in laboratories on Earth are negligible, and that a resolution of the measurement problem must be sought elsewhere.

Finally, there are approaches that doubt the validity of QM—notably, those based on hidden variables (Chapter 3). One of the tests of such approaches is the challenge they pose to the utter randomness of measurement results according to QM: any deviation from such randomness will signify the breakdown of the collapse or projection postulates and manifest hidden correlations with the apparatus. Thus far, no such deviation has been detected. On the contrary, this randomness has been confirmed with spectacular precision. Thus, to the probable displeasure of Einstein had he been witness to the unshakeable confirmations of QM at present, it is safe to state that God does play dice.

4.3 PARALLEL EVOLUTIONS (AND UNIVERSES?)

Born's superposition principle in QM appears to be in conflict not only with our daily-life experiences and sensual perception, but also with results of well-controlled physical experiments. There is no sphere of our experience where we directly observe a superposition of two or more eigenstates, but rather one eigenstate at a time. The evidence that quantum superposition states exist is always statistical, based on the analysis of data extracted from many measurements, as explained in Section 4.2.

There is yet another difficulty posed by the apparent dichotomy between Von Neumann's projection postulate and Born's superposition principle. How can there be two different realities—a quantum-mechanical one prior to a measurement, and a randomly chosen classical one thereafter?

Von Neumann and his close friend Wigner attributed this abrupt change in reality to a projection (or wavefunction collapse) that occurs in the observer's mind. Their view smacks of subjectivity and is similar to Bishop Berkeley's denial of objective phenomena (see Section 7.3). To paraphrase Berkeley, their view implies that if an observer does not look at the record of the measuring apparatus, the measurement has not occurred. It is impossible to accept this view in our automated world where all measurements are digitally recorded and processed without human intervention.

Here we dwell on another approach to this fundamental issue: the *many-world or parallel-universe interpretation* by H. Everett III (Princeton, 1957). This interpretation, which has become trendy of late, posits that each eigenstate in a superposition exists in another universe, so that in a given universe there is a unique albeit unpredictable outcome of any measurement; i.e., only one eigenstate is recorded each time. The typical evolution of a quantum system is a succession of an enormous number of elementary interactions and measurement events; e.g., collisions of the quantum system with photons or cosmic-ray particles hitting it. These particles may become correlated or entangled with the system, so that each of them effectively measures the system. In any time interval there is therefore a staggering number of (unread) measurement results and therefore of parallel universes: interactions with the environment keep splitting the universe at a mind-boggling rate. Everett's view is equivalent to the assumption (promoted by J. Preskill, Caltech, to this day) of an equally enormous variety of alternative histories, each generated by a different sequence of elementary events.

Clearly, this outlandish view does not abide by "Occam's razor" principle: it is a far cry from a minimalistic description of the world. Yet it poses another problem concerning the observer's role. If a measurement of a superposition state is performed by an observer, then the different results obtained in separate universes correspond to alternative versions of the observer: namely, *by splitting the universe the observer splits himself or herself.* Furthermore, if many observe the same event, each of them splits the other observers in his/her own fashion. There is thus no clear answer to the question: *Is there an objective (observer-independent) reality in the many-world view?*

Everett was not the first to express this odd view. Strikingly, the celebrated Argentinian writer and thinker J. L. Borges, though he never actively engaged in science, came to similar conclusions, probably by sheer intuition. In his thriller *The Garden of Bifurcating Paths* (1941) he wrote: "The Garden of Bifurcating Paths is an incomplete, but not false, image of the universe. A web of bifurcating, intersecting or mutually avoiding times ... covers all possibilities. In most of these times we do not exist; in some you exist but I don't; in others, only I but not you; in yet other times—both of us."

It is remarkable that contemporary cosmology speculates on the equally outlandish possibility of intersections among alternative histories in the so-called *multiverse*; i.e. the collection of all possible worlds. The proposed mechanism of such intersections is conjectural at this point, but suffice it to assume that alternative histories are somehow able to affect each other. If we ever come to verify this hypothesis, our perception of the flow of time may be completely reshaped—but so far our experience has always been causal. Cause always precedes and triggers effect, but according to the above hypothesis we may be able to experience child hood, adolescence and maturity in any chosen order, or vanish and be reborn at will by hopping between such histories without causal constraints. Such super-human existence will supersede all patterns of human existence and thinking which we have known so far. And in hindsight, the roots of this envisioned radical departure from causal, chronologically ordered thinking and existence may well be recognized to have started with Born's superposition principle.

Active observer

When you observe the quantum world
You bring about its demise.
But then, once more, lo and behold,
Another world does arise!

Observer, don't presume that you
Idly perceive how all flows by.
The world keeps being formed anew
Because it's watched by you and I.

4.4 APPENDIX: PROJECTOR OPERATORS

In this appendix we will further elucidate the notion of probability amplitudes, introduce new notations for operators, and present the projection operators.

In the previous appendices we introduced an unexplained normalization criterion for the probability amplitudes of a superposed state. Specifically, Henry's superposed sitting and leaning states were formalized as

$$\psi = a_S \psi_S + a_L \psi_L$$

with the normalization $|a_S|^2 + |a_L|^2 = 1$. We can now explain where this normalization comes from: $|a_S|^2 = p(S)$ is the probability to measure Henry in the sitting state, whereas $|a_L|^2 = p(L)$ is the probability to measure Henry in the leaning state. Their sum must equal 1, since there is a single Henry, so that when someone measures his state he must appear somewhere: The existence of a *single object* (Henry) is ensured by this normalization.

In order to mathematically describe a measurement we present a new formalism for operators. Until now, operators (e.g., Split and Recombine operators used by Henry) have been represented by matrices. Another way to represent operations is by the bra-ket formalism. Let us rewrite Henry's state in this formalism:

$$\psi = a_S \, | \, S \rangle + a_L \, | \, L \rangle$$

Here the state is described by kets $|S\rangle, |L\rangle$ which represent the sitting and leaning states, respectively. There is also a bra *conjugate* of each ket state denoted by $\langle S|, \langle L|$. In the matrix formalism:

$$|S\rangle = \begin{pmatrix} 1 \\ 0 \end{pmatrix}$$

$$\langle S| = |S\rangle^\dagger = \begin{pmatrix} 1 \\ 0 \end{pmatrix}^\dagger = \begin{pmatrix} 1 \\ 0 \end{pmatrix}^{T*} = \begin{pmatrix} 1 & 0 \end{pmatrix}^* = \begin{pmatrix} 1 & 0 \end{pmatrix}$$

where we have introduced the "dagger" symbol for the conjugate transpose of a matrix: transpose means a change of columns to row-vectors and vice versa. Thus, the bra of a state is its row-vector representation, whereas the ket of a state is the column-vector representation of the same state.

The bra-ket products of Henry's states are:

$$\langle S|S \rangle = \begin{pmatrix} 1 & 0 \end{pmatrix} \begin{pmatrix} 1 \\ 0 \end{pmatrix} = 1 \times 1 + 0 \times 0 = 1$$

$$\langle S|L \rangle = \begin{pmatrix} 1 & 0 \end{pmatrix} \begin{pmatrix} 0 \\ 1 \end{pmatrix} = 1 \times 0 + 0 \times 1 = 0$$

$$\langle L|S \rangle = \begin{pmatrix} 0 & 1 \end{pmatrix} \begin{pmatrix} 1 \\ 0 \end{pmatrix} = 0 \times 1 + 1 \times 0 = 0$$

$$\langle L|L \rangle = \begin{pmatrix} 0 & 1 \end{pmatrix} \begin{pmatrix} 0 \\ 1 \end{pmatrix} = 0 \times 0 + 1 \times 1 = 1$$

As can be seen, a bra-ket product of the same state is equal to 1, whereas a bra-ket product of two orthogonal (L and S) states is equal to 0.

Let us consider instead the corresponding ket-bra products:

$$|S\rangle\langle S| = \begin{pmatrix} 1 \\ 0 \end{pmatrix} \begin{pmatrix} 1 & 0 \end{pmatrix} = \begin{pmatrix} 1 & 0 \\ 0 & 0 \end{pmatrix}$$

$$|S\rangle\langle L| = \begin{pmatrix} 1 \\ 0 \end{pmatrix} \begin{pmatrix} 0 & 1 \end{pmatrix} = \begin{pmatrix} 0 & 1 \\ 1 & 0 \end{pmatrix}$$

$$|L\rangle\langle S| = \begin{pmatrix} 0 \\ 1 \end{pmatrix} \begin{pmatrix} 1 & 0 \end{pmatrix} = \begin{pmatrix} 0 & 1 \\ 1 & 0 \end{pmatrix}$$

$$|L\rangle\langle L| = \begin{pmatrix} 0 \\ 1 \end{pmatrix} \begin{pmatrix} 0 & 1 \end{pmatrix} = \begin{pmatrix} 0 & 0 \\ 0 & 1 \end{pmatrix}$$

Ket-bra products are *operators*—matrices that act upon states, changing them to other states. A special kind of ket-bra product are *projection operators* that represent measurements. In our case:

$$|S\rangle\langle S|\psi = a_S|S\rangle \langle S|S\rangle + a_L|S\rangle \langle S|L\rangle = a_S|S\rangle$$

$$|L\rangle\langle L|\psi = a_S|L\rangle \langle L|S\rangle + a_L|L\rangle \langle L|L\rangle = a_L|L\rangle$$

As can be seen, the $|S\rangle\langle S|, |L\rangle\langle L|$ operators *project* the state onto the S and L states, respectively. Regardless of the initial superposition state, the resulting state after a projection operator is the projected state.

We may now calculate the probabilities and amplitudes of the relevant states in this formalism. First, consider the state normalization:

$$|\psi|^2 = \psi^\dagger\psi = (\langle S| a_S^* + \langle L| a_L^*)(a_S|S\rangle + a_L|L\rangle)$$
$$= a_S^*a_S \langle S|S\rangle + a_S^*a_L \langle S|L\rangle + a_L^*a_S \langle L|S\rangle + a_L^*a_L \langle L|L\rangle$$
$$= |a_S|^2 + |a_L|^2 = 1$$

In the first line is the definition of $|\cdot|^2$ as the multiplication of a state with its dagger counterpart, and in the second line is the bra-ket notation introduced previously. Thus, multiplying a state by its dagger counterpart gives its probability. Now let us calculate the probabilities of collapsing to L or S by means of the projection operators:

$$p(S) = \||S\rangle\langle S|\psi\|^2 = |a_S|S\rangle|^2 = \langle S|a_S^*a_S|S\rangle = |a_S|^2 \langle S|S\rangle = |a_S|^2$$
$$p(L) = \||L\rangle\langle L|\psi\|^2 = |a_L|L\rangle|^2 = \langle L|a_L^*a_L|L\rangle = |a_L|^2 \langle L|L\rangle = |a_L|^2$$

Thus the probability of collapsing to a specific state is calculated by projecting the superposed state onto the collapsed state and taking its absolute-value squared.

Finally, we ask: what is the state *after* the collapse? Since the projection is an operator, is this state simply the projected state? Namely:

$$\psi_{\text{projection}} = |S\rangle \langle S|\psi = a_S|S\rangle$$
$$|\psi_{\text{projection}}|^2 = |a_S|^2 \neq 1$$

The answer is negative, since the projected state is not normalized: its probability is not equal to 1. Hence, simply projecting the state is not enough, and a normalization procedure must ensue:

$$\psi_{\text{after collapse}} = \frac{\psi_{\text{projection}}}{\sqrt{|\psi_{\text{projection}}|^2}}$$

To summarize, a measurement is represented by the following two stages:

1. A *random* selection of the collapsed state, with the probability given by the absolute-value squared of the projected-state amplitude.

2. Normalization of the projected state after the collapse.

Henry Bar Grows Uncertain

I'm reminded of the proverbial Heisenberg microscope: detecting an electron in an atom by two consecutive photons.

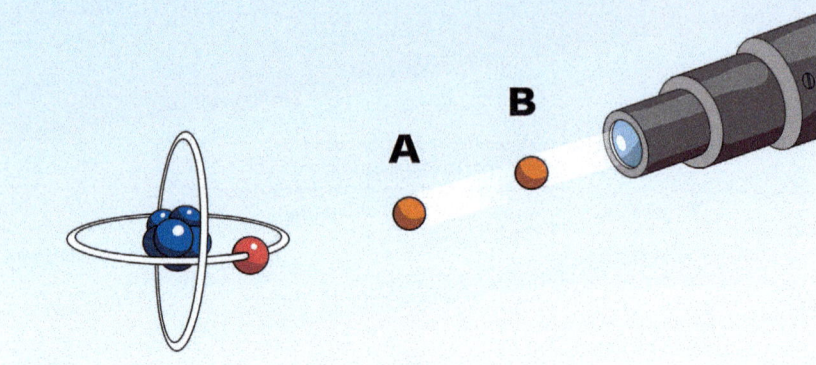

The first photon (A) pins the electron down, but increases its momentum uncertainty.

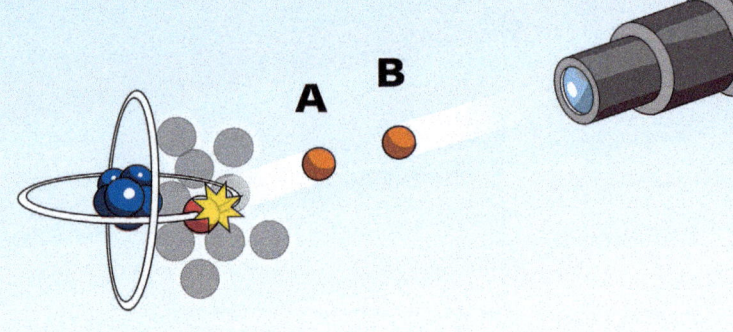

The next photon (B) can't find the electron because its position has become uncertain. That's what I did to baffle Eve!

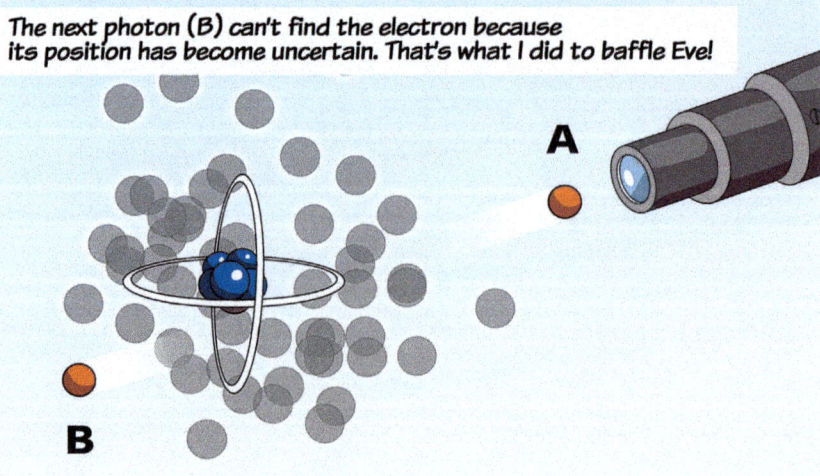

CHAPTER 5

What is Quantum Uncertainty?

5.1 HENRY BAR'S UNCERTAIN POSITION

In this adventure, Eve attempted to use her advanced (yet classical) devices to eavesdrop on Henry. But being a quantum object, Henry discovered that her eavesdropping strategy is susceptible to hindrances imposed by the Heisenberg uncertainty principle which is unavoidable in QM. As shown here, what makes her strategy vulnerable to Henry's quantumness is that Eve must perform two *consecutive measurements*—the first to pinpoint Henry, and the second to collect the sound waves of his conversations.

The crux of the matter is that as Eve pinpoints Henry's location, then, according to the quantum (Heisenberg) *uncertainty principle*, there is no restriction on the accuracy of her *single* measurement. Hence, Eve can detect Henry's whereabouts to any precision she chooses. However, this precision comes at a price: The uncertainty principle sets a limit to the *combined* accuracy of *two* consecutive measurements, so that if you perform the first one extremely accurately, you are bound to have a very uncertain result in the next measurement, provided some time has elapsed between the two. When Eve accurately locates Henry initially, this position measurement *induces large momentum uncertainty* in Henry's state. As a result, Henry acquires a large spread of velocities, so that, as time goes on, his subsequent position becomes extremely uncertain, defeating Eve's attempt to keep further track of his location.

When we consider such effects, we must keep in mind that whereas in previous chapters Henry's quantum states were superpositions of a discrete number of states, so that once measured, Henry's quantum superposition state collapsed to a *single* classical state, now, Henry's state is a superposition of his *continuous-observable* states, these observables being position and momentum.

Let us review the entire measurement process in this episode in light of the uncertainty principle. Henry first split into several versions of himself at different locations so that his position was "smeared" over a certain range. His state had a slightly uncertain position and slightly uncertain momentum, so that he was well-localized and was not moving much about that location. Next, Eve tried to measure his position to an extremely high accuracy. By doing so she *decreased* his position uncertainty. Now, Heisenberg's uncertainty principle (Chapter 2) states that position uncertainty and momentum uncertainty are inversely proportionate to each other. Hence, the drastic reduction in Henry's position uncertainty resulted in an *increase* in Henry's momentum uncertainty. Since momentum change produces a change in position in time, large momentum uncertainty means that Henry's subsequent position changed in an unknown way. Thus, Henry spread out very fast in many directions. When Eve next tried to tap Henry's conversation, assuming he was still at the location from where she had measured him just a short while earlier, she could not hear him. Henry was no longer there, or not only there. He was spread out too thinly for her to eavesdrop on him.

What is the origin of this counterintuitive uncertainty principle? One relatively simple explanation has to do with the measurement itself. A measurement has to be an *active process*; *there is no passive observation* in quantum mechanics (Chapter 4). In order to measure position, probes must be sent to interact with the measured object and return to the measuring device. For example, when we see an object it means that light probes—photons—travel from the light source, *interact* with the object, and only when they reach our eyes, which are our measurement devices, we actually detect them.

However, when probes, be they photons, electrons or billiard balls as Henry visualizes them, interact with the object, they change it. In Henry's case, some of the energy or momentum of the probe must be transferred to him (for he is the object). This momentum transfer changes the object's momentum in uncertain ways that cannot be predicted. The crux of the matter is that in order to pinpoint the object's *exact* location, one must send more and more probes. The more probes one sends, the more accurate or certain you are about the observable you measure, because each probe gives you a little more information about the position of the object. However, the more probes you send, the more momentum uncertainty they transfer to the object. Hence, there is an *inverse relation* between the uncertainty of the measured observable (here, position) that one reduces by repeated probe interactions and the uncertainty of the complementary observable (here, momentum) that one imparts by the same interactions.

Thus, Heisenberg's uncertainty principle states that there is a restriction (bound) on the maximum accuracy of *joint* measurements of two observables if they are deemed *complementary*. But which pairs of observables are complementary and thus obey this bound? For example, what if Eve wanted to measure Henry's position along two axes, say x and y? Could she do this with unlimited accuracy, or would the uncertainty principle play its tricks on her? When measuring position along the x-axis, probes are sent along this axis, increasing the x-axis momentum uncertainty. However, this measurement has no effect on the y-axis uncertainty in either position or momentum, because in order to measure the y-axis position, probes must be sent along the latter axis. Thus, the measurement of observables along one axis does not affect the observables along the other. Hence, conceptually, there is no limit to the accuracy with which Eve can measure Henry's position along two different axes.

There is a general answer to the question: which pairs of quantum observables can be measured together without accuracy restrictions and which cannot? The answer depends on whether or not these observables commute; namely, whether the order of the two operators associated with the observables matters. If they do not commute, then when operator A and thereafter operator B act on the quantum state, the resulting state will be different than if B acts first and then A. For non-commuting operators, such as position and momentum, the combined uncertainty of the corresponding observables cannot be less than a certain value. This value, which expresses the effect of their ordering, is proportional to the familiar $\hbar$. This is an indication of their quantumness: any quantity that involves $\hbar$ is essentially and unequivocally quantum in nature, which also means that the ordering effect is extremely small and hard to measure. Still, our quantum hero, Henry Bar, does not carry his $\hbar$ emblem in vain: his quantum suit can operate under conditions where $\hbar$ effects may save him from Eve's schemes.

In Chapter 6, Henry explores another strange form of the uncertainty principle to pre-empt Eve's attack on his friends.

5.2 UNCERTAINTY AND COMPLEMENTARITY IN QUANTUM MEASUREMENT THEORY

W. Heisenberg's *uncertainty principle* (established in 1927) was generalized and renamed the *complementarity principle* by N. Bohr that same year. It soon became the highlight of the Copenhagen view that QM is essentially a measurement theory (Chapter 4). According to this principle, any quantum observable measured on a given statistical ensemble (according to M. Born's statistical interpretation of

QM) has a complementary counterpart on the same ensemble. Mathematically, the two complementary observables are represented by operators that do not commute; i.e., the order in which they act on the object matters. What consequence does their non-commutability have?

Pairs of complementary (non-commuting) operators, such as position and momentum, or spin operators along two orthogonal axes (e.g., x and z) cannot share the same eigenstate. For example, an object in a momentum eigenstate is necessarily in a superposition of position eigenstates, and vice versa. It is up to the observer to decide whether the apparatus is set to measure momentum, thereby forcing the object to be in a superposition of position eigenstates. Since only an eigenstate yields *identical* measurement results time and time again on a given ensemble, whereas a superposition gives rise to random results, the observer thus renders the position of the object *uncertain* by measuring its momentum. Heisenberg's uncertainty relation signifies that the statistical spread of measured values of an observable is necessarily at the expense of the spread of its complementary counterpart: the smaller the spread (uncertainty) of one, the larger that of the other. This uncertainty relation is quantum since the product of the uncertainties of any pair of complementary operators is proportional to $\hbar$.

Bohr deduced from Heisenberg's uncertainty relations the broader *complementarity principle* whereby there is a fundamental limit on the information acquisition from the same statistical ensemble on complementary observables: the more you know about one, the less you know about the other.

This restriction on the maximal accuracy of measuring physical quantities imposed by the mathematical structure of QM has no parallel in classical (Newtonian) mechanics. It therefore stirred stormy debates during the Solvay conferences between proponents of the Copenhagen interpretation and the self-proclaimed realists, headed by Einstein, who rejected any conceptual breach between quantumness and classicality. The debate concerning complementarity invoked "thought" (*gedanken* in German) experiments that both parties interpreted differently.

One such *gedanken* experiment was Heisenberg's microscope. He imagined a microscope that employs photons with extremely short wavelengths (currently known as gamma quanta) capable of accurately pinpointing the position of an electron orbiting the nucleus of a hydrogen atom. He then calculated that if a measurement is effected by such a photon that scatters the electron and provides the microscope with a sufficiently precise position of the electron (within the size of the atomic orbital, 0.05 nanometers), then an attempt to repeat this measurement may fail utterly. The reason is that the first highly accurate position measurement may produce a huge momentum change of unknown size and direction, according to the uncertainty principle. As a result of this change, the electron can acquire enough kinetic energy to be kicked out of its orbital and escape far from the atom. Of course, since the results are unknown (random), we may be lucky enough to detect the electron again, but if many measurements are performed on a given atom, such runaway effects are to be expected. Heisenberg's microscope pointed to a possible *destructive* nature of a quantum measurement, which may cause the object (an atomic electron) to cease to exist in the setup. However, it remained unclear whether the uncertainty principle applies to scenarios where no such devastation is caused by the measurement.

Einstein challenged the uncertainty principle by a *gedanken* two-slit setup for the electron wherein hitting the left-hand or right-hand slit is capable of making the plate slide left or right, by attaching it to springs and rolls (Figure 5.1). He considered an electron with well-defined momentum, which according to QM may pass through either slit without our knowing which. However, as the electron traverses the slit, it kicks it slightly, and the corresponding change in the plate position can be recorded. Yet the kick can be weak enough that the momentum does not change appreciably, allowing many electrons to produce an interference pattern on the screen far beyond the two-slit plate. The pattern allows an accurate determination of the momentum, and the position shift of the rolling plate can disclose which slit the electron traversed in each run. Einstein claimed that the combined uncertainty of such position and momentum measurements can be as small as one likes, in contradiction to their alleged complementarity.

And yet Bohr and his associate L. Rosenfeld could explicitly show the fallacy of this argument. They calculated that if a kick given by the electron to the rolling plate were large enough to record which slit was hit, it would necessarily create such a spread in the electron momentum that the interference pattern on the screen would be washed out. Their result reaffirmed Heisenberg's uncertainty principle, and Einstein was forced to agree with their conclusion, though he remained convinced that QM should be superseded by a more complete theory, free of complementarity constraints.

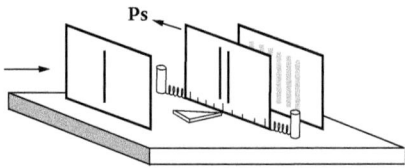

Fig. 5.1 Einstein's suggested setup in his controversy with Bohr on the validity of complementarity based on the Heisenberg uncertainty relation in two-slit interference. Does the position shift of the rolling plate disclose which-path (or which-slit) information and yet allow for an interference pattern?.

5.3 IS UNCERTAINTY HUMAN?

The Copenhagen school viewed the inability to acquire complete knowledge on complementary observables as a fundamental limitation on human perception; i.e., on epistemology. Bohr maintained that our cognition does not allow us a complete, comprehensive view of the world, which comprises a multitude of diverse attributes that are inherently complementary from a human perspective. By advancing this argument he essentially adopted (without acknowledging it) I. Kant's eighteenth-century view that we cannot perceive the world as it "really" is, it being a "thing in itself" (*ding an sich* in German). Instead, we "filter" it through our innate perception. Kant originally postulated that the unshakeable truth of Newtonian physics stemmed from the categories of space, time and causality that human thought and perception possess. By the same token, Bohr asserted that complementarity as revealed by the uncertainty relation is unavoidable, being the mental "filter" through which quantum phenomena are grasped by human observers. Bohr's acknowledged influence was the nineteenth-century Danish existentialist philosopher S. Kirkegaard. Like him, Bohr placed the human "self" at the center of the perceived universe and relinquished the possibility of objective knowledge.

In this respect there is a common denominator linking Bohr, Schrödinger and Von Neumann: they all insisted on the central role of human cognition and the supremacy of mind over matter in their interpretation of the role of the observer in QM. Yet only Bohr elevated complementarity and the associated uncertainty relation to the rank of fundamental truth.

Although philosophical controversy had accompanied "classical" physics for centuries, the intense debate on "the meaning of it all" in the first decade of quantum mechanics (QM) was unprecedented. It appeared that without the resolution of the underlying philosophical issues—primarily the role of the observer (Chapter 4)—QM made no sense. No wonder that the "realists" led by Einstein, who viewed the world and our knowledge of it as free of human bias, rejected Bohr's insistence on the inevitability of complementarity. Yet regardless of philosophical preference, no one can deny to this day the experimentally proven validity of the uncertainty relations and their consistency with the rules of QM. Einstein's grudging acknowledgment of his defeat in the controversy with Bohr and Rosenfeld (Section 5.2) marked the acceptance of Bohr's complementarity or the uncertainty principle by the consensual opinion of the physical community which has never been overturned.

It is still incumbent upon us to produce an explanation for complementarity better than Bohr's. One option is to deepen our insight into the observer's role in QM. By *treating the observer quantum mechanically*, as we must if QM is indeed universal, we have to allow for entanglement/correlations between the observer and the detected object. According to Von Neumann's quantum measurement theory, such entanglement or inseparability is a prerequisite for a measurement. If the observer and the observed object are in an inseparable state, the uncertainty of a quantum observable may reflect the averaging of this joint state over many indistinguishable states of the observer that are covertly correlated to the state of the observable—say, the mental or cognitive states of the observer. Conversely, each mental state of the observer may be correlated to many quantum observables. When the states of these quantum observables are ignored or cannot be discriminated by the observer, the observer's cognitive functions may perceive randomness or uncertainty of the physical data or information. Thus, a study of such surmised correlations, which is unfeasible at present, may in future shed new light on the origin of complementarity and uncertainty in QM.

Such an approach would be in the spirit of B. Spinoza's seventeenth-century monistic philosophy. Spinoza viewed nature as *one substance* with both mental and physical attributes or modalities whose correlations may be revealed by spiritual progress. Perhaps one day a comprehensive science that encompasses physics, information theory, biology and even psychology may accomplish Spinoza's program of viewing the world from its different aspects and free our knowledge from complementarity constraints.

Uncertain World

Bohr taught us: it is elementary
That measurements are complementary.
Peeping behind the world's curtain
We find our knowledge is uncertain.
Yet from here on thickens the plot:
Is such uncertainty our lot?
Or shall we one day free our mind
And bring true knowledge to mankind?

5.4 APPENDIX: CONTINUOUS VARIABLES

In this appendix we introduce the mathematical notions of operators and functions that act on continuous variables. The mathematics of continuous variables is more complex than that of discrete variables, so that we shall resort to certain relations without proof. Our aim is to explain the mathematical essence of the uncertainty principle.

Quantum states representing continuous variables such as position and momentum are described by *continuous functions* of these variables. Thus, for example, a quantum state of position is represented by $|\psi\rangle = \int_x dx f(x)|x\rangle$, normalized by $\int_x dx |f(x)|^2 = 1$. Here, $f(x)$ is the probability amplitude distribution over positions x, and $|f(x)|^2$ represents the probability distribution. Here we have introduced the integral sign, $\int_x dx$, which signifies a summation over x values inside infinitesimal (infinitely small) intervals dx.

Thus, for example, a particle localized around x_0 may have the bell-shaped Gaussian distribution (Figure 5.2) over position x in the form:

$$|f(x)|^2 = \frac{1}{\sqrt{2\pi \Delta x}} e^{-\frac{(x-x_0)^2}{2\Delta x^2}}.$$

whose uncertainty or width is measured by the standard deviation Δx. The first factor in this function accounts for its normalization. The next factor is the constant $e = 2.718281$, elevated to a power (here given by the squared term in the exponent) known as the exponential function. The squared term in the exponent is responsible for the bell shape of the function.

If Δx is small, position x is known with high accuracy to be near x_0, but if Δx is large, x can acquire a wide variety of different values (or be spread over a broad range of x) with high probability. This function is a generic (though not unique) representation of an uncertain position x.

The same state can be described in the momentum representation as:

$$\psi = \int_p dp \tilde{f}(p)|p\rangle,$$

with the normalization $\int_p dp |\tilde{f}(p)|^2 = 1$. Here, $\tilde{f}(p)$ represents the particle probability amplitude to have momentum p. The probability distribution $|\tilde{f}(p)|^2$ *must* be Gaussian if the position is Gaussian (the proof is left for a later chapter):

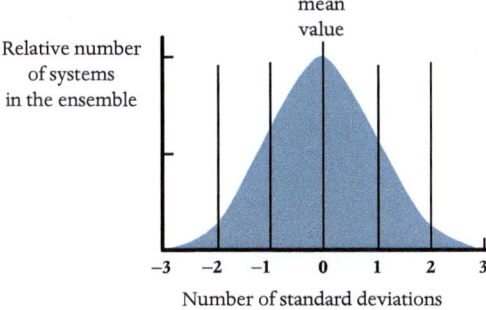

Relative number of systems in the ensemble

mean value

−3 −2 −1 0 1 2 3

Number of standard deviations from an average value

Fig. 5.2 A quantum state of a particle with Gaussian distribution of position x.

$$\left| \tilde{f}(p) \right|^2 = \frac{1}{\sqrt{2\pi \Delta p}} e^{-\frac{(p - p_0)^2}{2\Delta p^2}}$$

The question we next wish to answer is: what is the lower bound on the uncertainty of position and momentum?

Heisenberg's uncertainty principle relates the uncertainties of any two non-commuting operators in the following way:

$$\Delta A \Delta B \geq \frac{1}{2} \left| \langle [A, B] \rangle \right|$$

where $[A, B] = AB - BA$ is the commutation relation and $\langle O \rangle = \langle \psi | O | \psi \rangle$ is the expectation value of an operator. The larger-or-equal sign of the inequality means that the lower bound on the uncertainty product on the left-hand side is when the equality sign with the commutation relation on the right-hand side holds. But what is this bound explicitly?

At first glance, the commutation relation looks odd: if A and B were numbers it would be always 0. However, in quantum mechanics, observables such as A and B are not numbers but *operators*: their commutator is non-zero because the order in which the operators A and B act on a quantum state may matter a great deal.

Specifically, for position and momentum the uncertainty principle has the form:

$$\Delta \mathbf{x} \Delta \mathbf{p} \geq \frac{1}{2} \left| \langle \psi | (\mathbf{x}\mathbf{p} - \mathbf{p}\mathbf{x}) | \psi \rangle \right|$$

So far, we have treated x and p as variables, in terms of which a quantum state is represented by the functions $f(x)$ and $f(p)$. We now consider the meaning of these representations when we evaluate the uncertainties of the operators $\mathbf{x}$ and $\mathbf{p}$.

The position operator $\mathbf{x}$, in the position representation introduced previously, simply yields the *variable value* x for state $|x\rangle$. $\mathbf{x}$ is also called an eigenfunction with eigenvalue x. For the expectation value of x, we have the following calculation:

$$\mathbf{x} \mid x \rangle = x \mid x \rangle$$
$$\mathbf{x} \mid \psi \rangle = \int_x dx f(x) \mathbf{x} \mid x \rangle = \int_x dx\, x f(x) \mid x \rangle$$
$$\langle \mathbf{x} \rangle = \langle \psi | \mathbf{x} | \psi \rangle = \iint_{x,x'} dx dx'\, \langle x' | f(x')\, \mathbf{x} f(x) | x \rangle$$
$$= \iint_{x,x'} dx dx'\, \langle x' | f(x')\, x f(x) | x \rangle = \iint_{x,x'} dx dx' f(x')\, x f(x)\, \langle x' | x \rangle$$
$$= \int_x dx | f(x) |^2 x$$

The first line represents the operator $\mathbf{x}$, the second line shows its operation on a general wavefunction, and the third and fourth lines are the calculation of the expectation value of $\mathbf{x}$. In the transition to the last line we used the fact that the states $| x \rangle$ form an orthonormal basis:

$$\langle x'|x\rangle = \begin{cases} 1 & x' = x \\ 0 & x' \neq x \end{cases}$$

As can be seen, the expectation value of $\mathbf{x}$ is the weighted average of the position, since $|f(x)|^2$ represent the probability of position x.

For the momentum operator, $\mathbf{p}$, we can follow exactly the same calculation in the momentum representation, but that will not help us to calculate the commutation relation, since both $\mathbf{p}$ and $\mathbf{x}$ must be calculated in *the same representation*. Hence, we must know what the momentum operator $\mathbf{p}$ looks like in the position representation.

Without giving any proof, we present the momentum operator in position representation in the one-dimensional form (the derivation will be shown in later chapters):

$$\mathbf{p} = \frac{\hbar}{i}\frac{d}{dx}$$

Thus, the momentum operator in the position representation involves the following factors. First, the already familiar $\hbar$; and second, the letter i, which represents $i = \sqrt{-1}$, also called the imaginary (more on this in subsequent chapters). The last factor is called the *derivative with respect to x*, representing the change with respect to x of any function that follows it.

To fully explain the uncertainty relation, we only need two extremely simple characteristics of the derivative:

(i) $\frac{d}{dx}x = 1$; in other words, the linear function x changes with respect to x by unity.

(ii) The chain rule: $\frac{d}{dx}\left(f(x) \times g(x)\right) = \frac{df(x)}{dx}g(x) + f(x)\frac{dg(x)}{dx}$; this rule means that the derivative of a product of functions is the sum of terms in which the derivative acts on each function in its turn, but only once.

We are now able to calculate the uncertainty relation bound as follows:

$$\langle[\mathbf{x},\mathbf{p}]\rangle = \langle\psi|\,[\mathbf{x},\mathbf{p}]\,|\psi\rangle$$

$$= \iint dx dx'\, \langle x'|f(x')\,[\mathbf{x},\mathbf{p}]f(x)|x\rangle = \iint dx dx'\, \langle x'|f(x')\,(\mathbf{xp}-\mathbf{px})f(x)|x\rangle$$

$$= \iint dx dx'\, \langle x'|f(x')\,\mathbf{xp}f(x)|x\rangle - \iint dx dx'\, \langle x'|f(x')\,\mathbf{px}f(x)|x\rangle$$

$$= \frac{\hbar}{i}\iint dx dx' f(x')\,\frac{df(x)}{dx}x\,\langle x'|x\rangle - \frac{\hbar}{i}\iint dx dx' f(x')\,\langle x'|\frac{d}{dx}f(x)x|x\rangle$$

$$= \frac{\hbar}{i}\iint dx dx' f(x')\,\frac{df(x)}{dx}x\,\langle x'|x\rangle - \frac{\hbar}{i}\iint dx dx' f(x')\,\left\langle x'|\left(\frac{df(x)}{dx}x+f(x)\frac{dx}{dx}\right)|x\right\rangle$$

$$= \frac{\hbar}{i}\iint dx dx' f(x')\,\frac{df(x)}{dx}x\,\langle x'|x\rangle - \frac{\hbar}{i}\iint dx dx' f(x')\left(\frac{df(x)}{dx}x+f(x)\frac{dx}{dx}\right)\langle x'|x\rangle$$

$$= \frac{\hbar}{i}\int_x dx\left(f(x)\frac{df(x)}{dx}x - f(x)\frac{df(x)}{dx}x - |f(x)^2|\right)$$

$$= -\frac{\hbar}{i} = i\hbar$$

Here:

- In the second line we used the commutation relation.
- In the fourth line we used the momentum operator definition.
- In the fifth line we used the chain rule.
- In the seventh line we used the orthonormality of the x basis.
- In the last line we used the normalization of the probability amplitude.

Hence, the uncertainty principle for position and momentum is:

$$\Delta x \Delta p \geq \frac{\hbar}{2}$$

We have changed the imaginary number i to its $|i| = 1$ absolute value according to the uncertainty principle formula.

This uncertainty relation means the following. If, say, the position uncertainty is reduced by a factor of 2, then the uncertainty of the momentum must increase by a factor of 2 because there is a minimal bound on the joint uncertainty of the position, and the momentum is equal to $\hbar$ divided by 2.

In this appendix we have introduced continuous variables, operators and their commutation relation in order to clarify the position–momentum uncertainty relation.

Henry Bar's Uncertain Jumps

CHAPTER 6

What is Time–Energy Uncertainty?

6.1 THE QUANTUM ROCKET AND TIME–ENERGY UNCERTAINTY

In the present episode, Henry scores a surprise win over Eve thanks to the new gadget he has worked hard to secretly integrate into his quantum suit: a quantum rocket. This new contraption is composed of a unique quantum-chargeable battery and a jet booster that utilizes the battery energy for high jumps. The new gadget can act in a counterintuitive, *non-classical* manner, as Henry finds out, much to his relief.

In Chapter 5 we introduced the uncertainty principle for position and momentum, whereby the more precise the position measurement, the less precise will be a subsequent momentum measurement. Is Henry's quantum rocket subject to an analogous uncertainty relation? To understand this issue, consider that the quantum rocket operates in two consecutive steps. In the first one, the batteries are charged with a certain amount of energy. In the second step, energy from the charged battery is released in the form of jet-engine work. The first step takes some time, as the charging is not instantaneous. During the second step, a "measurement" of energy takes place: the more energy is "measured", the higher Henry jumps. The question then arises: what "observable" is measured during the first step whose uncertainty is complementary to energy? Intriguingly, the first "observable" is the *time* interval over which Henry charges the battery. Thus, the uncertainty principle here involves *time* and *energy*.

Yet, as discussed in Sections 6.2 and 6.3, *time is not a quantum observable*, hence there is *no commutation relation between time and energy*. How, then, can *time* be uncertain and affect the energy uncertainty in the battery?

In order to answer this question, one must first define how one measures time in a quantum system. In Henry's quantum rocket, the quantum system is represented by the battery. As opposed to a classical battery, which can be *either* empty *or* charged, the quantum battery can be in a *superposition* of different energy states. Until one measures the energy, the battery can have *any* energy value. This uniquely quantum scenario results in an energy uncertainty of the battery. Time, on the other hand, is related to the *duration* of the quantum-state evolution. Specifically, a short charging burst results in higher energy uncertainty as compared to slower, steady charging that yields a known, well-defined energy state. This is the essence of the time–energy uncertainty: the shorter the process, the less we know about the energy; i.e. the larger the energy uncertainty. We note that shorter duration does not mean less *average* energy: the same amount of average energy is transferred, regardless of the duration, and only its statistical spread (variance) changes.

Let us recapitulate on Henry's latest adventure. In his first race against Alice and Bob up the Pisa tower, Henry charged his quantum-jet battery over a long time. This *time-uncertain charging process* resulted in a known energy state of the battery; i.e. in a pre-determined charging energy with a very low *energy uncertainty*. Henry was certain that he had the battery fully charged and was ready to jump to the tower level they had agreed on. During the second race, Henry saw that Eve was poised to abduct Alice. Since there was no time to lose, he had to charge his battery extremely fast. The charging transferred the same *average amount of energy*, but the short duration resulted in an extremely high *energy uncertainty*. Henry's quantum battery was in a superposed state of many charged energies; both below and above the average. This quantum effect may result in an extremely bizarre situation, wherein the battery has *more energy than the average charging energy*. While this is unlikely—that is has low probability—it may still happen. Thus, when Henry pressed the Discharge button, the *energy measurement* "collapsed" (Chapter 4) the battery to the state which landed him squarely at the top of the tower, between Alice and Eve, allowing him to pre-empt the abduction. This collapse, as in any quantum measurement (Chapter 4), could have ended up randomly in any of the available energy states. Henry had the good fortune to collapse to the most favorable one that allowed the highest jump. Yet, being a quantum physicist, Henry knows better than to keep relying on luck, as we shall see.

6.2 THE TIME–ENERGY UNCERTAINTY RELATION IN QM

Shortly after the momentum–position uncertainty relation was presented by Heisenberg and elevated to the rank of the fundamental complementarity principle (Chapter 5), the time–energy uncertainty relation was put forward by Bohr. The crux of both relations appeared to be similar. Just as with position and momentum measurements (Section 5.2), the precision of measuring (or knowing) energy is the larger, the smaller the precision of its time measurements (on a given ensemble of quantum systems).

However, there also seemed to be a clear difference between the two relations. Whereas the position–momentum uncertainty reflects the non-commutative nature of the two operators, this is not the case for time and energy, because *time cannot be described by an operator* (Section 6.3). In fact, anyone as well versed in Maxwell's classical electromagnetic (EM) wave theory as the pioneers of QM could easily recognize this relation to be the same one satisfied by a classical EM pulse of any finite duration: the shorter the pulse, the less can we characterize its frequency, and vice versa. And since (Chapter 1) in a quantized EM field the energy of photons is equal to their frequency times $\hbar$, at least for photons the Heisenberg relation is the same as for classical waves. De Broglie's relation between the energy and momentum of a massive quantum particle (Chapter 2) provided a straightforward, not surprising, analogy of the classical-wave time–energy uncertainty relation to its quantum counterpart because of its wavelike nature.

Yet Bohr and Heisenberg insisted on an equally fundamental status for the position–momentum and time–energy uncertainty relations. As with other basic issues of QM, battles raged on at the Solvay conferences (Section 5.2). Einstein—Bohr's contrary spirit—challenged the universality of the time–energy uncertainty by proposing a setup (a *gedanken* experiment) where wavelike properties associated with position–momentum uncertainty appear to play no role whatsoever. The setup consisted of a box (a cavity) enclosing an atom, and the cavity is hooked to a sensitive balance (scale) and has a shutter connected to a precise (macroscopic) clock (Figure 6.1). The atom is prepared in an excited state, whereupon the shutter is closed abruptly and the cavity is weighed. After a precisely known time interval the shutter is reopened and the cavity is weighed again. If the weight has decreased, we then know that the photon has been emitted within that time interval, and has changed the atomic mass according to Einstein's formula for the connection between the energy (carried off by

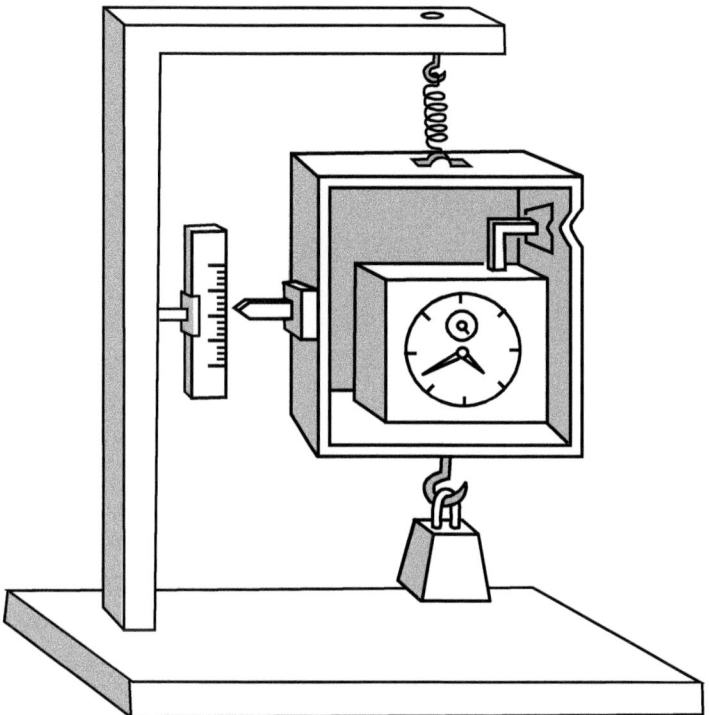

Fig. 6.1 Einstein's proposed setup for challenging the time–energy uncertainty relation. The Bohr–Rosenfeld rebuttal ended the controversy on the uncertainty relations in QM (see text).

the photon) and the mass. Einstein's argument was that there is no connection between the precision of measuring the energy change and the time interval, so that the product of their uncertainties or imprecisions should not be bounded by $\hbar$, as proclaimed by the Bohr–Heisenberg uncertainty relation.

After a sleepless night of pondering over Einstein's challenge, Bohr and his associate Rosenfeld came up with the following rebuttal. They pointed out that by Einstein's general theory of relativity the time interval is subject to dilation which is caused by the gravitational force used to weigh the atom. This force is in turn related to the momentum change of the atom that recoils as it emits the photon over this time interval. On the other hand, the position change of the scale spring determines the precision of the weighing. Consequently, the time-dilation and energy uncertainties can be translated into position–momentum uncertainties. By this reasoning, the time and energy uncertainty product has the same bound as the momentum–position uncertainty product which satisfies the Heisenberg uncertainty relation. Since Einstein had already accepted the latter uncertainty (albeit reluctantly, Section 5.2) and Bohr's other argument invoked his general relativity, Einstein had no choice but to admit defeat again.

6.3 TIME AND ENERGY IN OUR QUANTUM WORLD

The discussion in Sections 6.1 and 6.2 raises the question: what is the meaning of time, energy and their uncertainty in physics and in human experience in general?

Let us first muse on time—perhaps the most enigmatic notion of our existence, but also the one we are most acutely aware of. It is deeply rooted in our intuition (Kant maintained that it is an innate category of our thinking, one we are born with): everyone understands what the Queen of Hearts (in *Alice in Wonderland*) meant by "He is killing the time". Yet any attempt to define time formally opens a Pandora's Box of ambiguities and possibly inconsistencies.

The very *reality of time* has been debated over millenia. The first recorded controversy on this issue was between two schools of Ancient Greek philosophy. Heraclitus asserted that the world is changing incessantly, so that time consists of fragmented instants. By contrast, Parmenides rejected the reality of change and therefore time. His disciple, Zeno of Elea, put forward his paradoxes (the race of Achilles and the turtle, or the flight of an arrow) that were aimed at showing the inconsistency of the concept of motion as a manifestation of time (to be discussed in Chapter 10).

The founders of the leading schools of Greek philosophy, Plato and Aristotle, rejected the extreme views of both Heraclitus and Parmenides concerning time, instead advancing their own complex view: they discerned between eternal ideas (Plato) or eternal forms (Aristotle) to which time does not apply, and ephemeral matter that changes with time. This dual attitude to time later became part of medieval philosophy, culminating with Spinoza, who regarded time as a fleeting, insignificant mode of our existence.

The birth of modern physics with Galileo and Newton in the seventeenth century brought about the need for counting time in order to characterize motion, be it uniform (at a constant velocity) or accelerated. Remarkably, the Dutch physicist Huygens discovered in 1650 the synchronization of pendulum clocks that hang on the same wall and couple to each other, albeit weakly, via the wall. From then on, until the twentieth century, mechanical clocks— essentially periodic oscillators—served as time standards. They were presumed to count *universal time*, which does not depend on the motion of the observer or the observed object.

The assumption that time is universal was overturned by Einstein's special relativity (1905), in which the remarkable effect of time dilation arises: the faster the motion of a clock relative to the observer, the slower it will tick in the observer's eyes. A spectacular confirmation of time dilation was given in the 1930s by P. W. Anderson's detection of unstable elementary particles that cross a detector at a velocity close to the speed of light. Because of their high velocity, their internal "clock" that counts their lifetime slows down drastically, and the particles live much longer (on average).

Einstein's general relativity (1916) gave rise to the emergence of relativistic cosmology which recounts the "history of time" in the universe since the Big Bang (see S. Hawking's *A Brief History of Time*). It describes the time evolution on cosmic scales, on the assumption that the initial state of the universe set its complex *cosmological clockwork* in motion.

Another kind of temporal evolution was introduced in the late nineteenth century by Boltzmann and Gibbs, who combined thermodynamics with statistical physics and came up with the fundamental insight that typical processes involving heat exchange tend to increase the disorder of atoms in a macroscopic system. As disorder (associated with the notion of entropy) increases or ordered states decay, our information on the system decreases. According to Boltzmann and Gibbs, the "arrow of time" is directed towards the future, where less information is available on typical systems than in the past or present. According to this view, there is a thermodynamic, information-related clock in the universe.

Quantum mechanics (QM) has inherited all the foregoing notions concerning time: its measurement by oscillators, cosmological evolution or entropy increase associated with quantum-state decay. However, QM has an additional ingredient: time–energy uncertainty, which stems from the nature of the Schrödinger-wavefunction evolution. Since a wave extends over all space and time, its energy is precisely known only if we observe it everywhere and at all times, which is never the case. The more limited the time interval of observing the wavefunction, the less precisely we may determine its energy.

There appears to be an inconsistency in the very attempt to introduce time-dependence into the description of the universe by QM, as changes that make the *cosmological clock* "tick" must stem from an inherent non-stationarity or instability of the *entire* universe, which can only have its origin *outside the universe*. But is not the universe "everything" by definition? Equivalently, the universe would not conserve energy if it evolved as a whole. This inconsistency may be avoided if time-dependence only involves *parts of the universe* and occurs through their interaction with other parts, so that energy is neither conserved nor precisely determined within each such part, consistent with time–energy uncertainty. But then, can we talk of a unique cosmological clock? This issue remains enigmatic, despite attempts to resolve it conclusively within the QM framework.

An important caveat is in order. There is no requirement for all three types of clocks mentioned above to be subject to the *same time–energy uncertainty*: a given measurement may affect one kind of clock, but not another:

1. Bohr's assertion that an atom enclosed in a box may undergo time dilation under a gravitational force is consistent with the ultrahigh sensitivity of atomic clocks to gravity variation and with the need to correct GPS clocks for gravitational time dilation.

2. Thermodynamic and oscillator clocks are highly pertinent to quantum time-evolution, as may be seen for an atom that is prepared in its excited state. Its excited state then starts decaying through the emission of a photon. Only after a sufficiently long time can we be certain that the atom has decayed to its ground state and that a photon has been emitted with an energy that is the same as the energy difference between the excited and ground states of the atom, according to the law of energy conservation. Hence, at long times the energy of both the atom and the photon is precisely known, but not so at much shorter times, when neither the atom nor the photon have well-defined energies and it is impossible to determine whether the atom has decayed or not. During that short time we may view the atom and the photon as coupled oscillators exchanging energy back and forth, somewhat analogous to the (classical) coupled pendulum clocks studied by Huygens in the seventeenth century. The difference is that in the quantum decay process none of the oscillators has a well-defined energy or phase of oscillation, so that there is less information available on each of them than initially: Namely, entropy has increased in the course of the decay process, which means that the thermodynamic clock has been set in motion. Even after the emission has been completed, the information is not fully recovered because the photon may occupy any of the infinitely many modes available to it in free space.

The foregoing discussion suggests that QM has thus far adhered to similar notions of temporal evolution or time-measuring clocks as do classical physics or thermodynamics. This adherence is a consequence of our inability to treat time as a quantum observable (or operator). All attempts for such treatment fail, because if time were an observable whose values range from the infinitely remote past to the infinitely remote future, then the complementary quantum observable would have values that range from infinitely negative energy (which *does not exist* in the universe) to infinitely positive energy. Because of the unphysical nature of negative energy, the complementary time operator must be abandoned.

Yet there is a remarkable consequence of time–energy uncertainty in QM that has no classical counterparts: the notion *of virtual quanta*. According to energy conservation, such quanta are not supposed to exist. For example, if we have an unexcited atom in an empty cavity, then they will remain in this state over long times. But if we probe the atom or the cavity at very short time intervals, we may find, contrary to our classical intuition, that the atom is excited and/or there are photons in the cavity. The reason for this "miraculous" emergence of quanta or excitation "out of nothingness" (*ex nihil*, as medieval philosophers would have said) is that energy is not well-defined over short time intervals. More generally, one may view any empty volume of space (the vacuum) as swarming with incessantly emerging and almost immediately disappearing virtual quanta, only because the energy in this volume may randomly fluctuate over very short times.

Even more intriguingly, two unexcited atoms may interact with each other by exchanging virtual quanta on very short time-scales. The paradox is that even after averaging their interaction over long times, their total energy is changed compared to what it would be without these virtual quanta. Such change in the two-atom total energy produces an attractive force between the atoms known as the Casimir vacuum force, whose form at short distances between atoms is known as the van der Waals force. These *forces are of purely quantum-mechanical origin*.

But may we expect at some point a more drastic revision of the concept of time that would reflect the quantum-mechanical laws of nature? At present there is at least one direction of research that may bring about such revision: the *quantization of space and time* at intervals that are small enough to localize a quantum particle by its interaction with its own gravitational force. The length of such an interval (known as the Planck length) divided by the speed of light is known as the Planck time: 5.4×10^{-44} sec. Such exceedingly short intervals cannot be probed at present, yet they have a potentially great significance: their existence would mean that *time in QM is not continuous*, as it has been assumed thus far in physics, but rather *discrete*. This intriguing possibility is reminiscent of the view Heraclitus of Ephessos adhered to—that time is a collection of fragmented instants, so that the world arises anew at each consecutive instant. But we have a long way to go before we can confirm or reject this conclusion.

The quantum clock

Devouring time, thou that flies so fast,
Buries or sweeps away the past,
Can we be freed from thy fierce power,
Escape the unforgiving hour?
By quantum wisdom that is so concise,
Time's uncertain if energy's precise,
Thus we can be both young and old, but hear before ye rave:
To spread in time as ye dream, ye first must be a wave.

6.4 APPENDIX: FINITE-TIME EVOLUTION

In order to mathematically explain and formulate the time–energy uncertainty in Henry's latest adventure, we will first describe the quantum system—the battery—and then introduce the Schrödinger equation that governs the dynamics of quantum systems at finite times.

The quantum system is composed of several energy states, each with its own energy, represented by E_i, where $i = 0, \ldots, N$. These energies may take any values ranging from 0—the energy of the initial uncharged state—up to some maximal energy of the fully charged battery. However, unlike their classical counterparts, the battery can be in a *superposition* of quantum energy states, with probability amplitudes a_i (Section 4.4, Appendix):

$$|\psi> = \sum_{i=0}^{N} a_i |E_i>$$

In this formulation, $p_i = |a_i|^2$ is the probability that the battery will be charged to the E_i energy. Hence, at the end of the charging process the *average energy* of the battery is given by $\sum_{i=0}^{N} p_i E_i = E$, which we here take to be the *same as the total energy*.

However, the charging process is dynamic; it takes time, hence the probability amplitudes of the superposed quantum (energy) states *change with time*, depending on the charging process. Let us describe the *charging process* of the battery, which is governed by the duration of the process. A square function will represent a *constant flow (steady rate) of energy* throughout the charging process:

$$f(t) = \begin{cases} E/\tau & 0 < t \leq \tau \\ 0 & otherwise \end{cases}$$, where E is the average charging energy and τ is the charging duration.

How can we calculate the entire amount of energy transferred during the charging? We may sum up the energy flow over the duration, but how many time-points should we take? Time is continuous, so that we cannot write $\sum_{t=0}^{\tau} f(t)$, since a sum is over a *discrete* number of steps. For this purpose we use the *integral* introduced in Chapter 5, which is akin to a sum but is defined for continuous variables:

$$\int_0^\tau dt f(t).$$

Here, $\int$ is the integral sign: in it, the lower number represents the time at which we start (here 0), the upper number represents the time at which we stop (here τ), and dt represents the elementary (infinitesimally small) step we are summing over. If we insert in the integral the constant-flow function that Henry uses, we obtain:

$$\frac{\int_0^\tau dt f(t) = \int_0^\tau \frac{dt E}{\tau} = E}{\tau \int_0^\tau dt = \left(\frac{E}{\tau}\right) \tau = E}.$$

Here we used the fact that $\int_0^\tau dt = (\tau - 0)$, i.e. an "empty" integral is just the difference between its two limits. Thus, the total energy transferred over the full duration of the process is $\left(\frac{E}{\tau}\right) \tau = E$. As noted previously, the time uncertainty is equal to the *battery charging* duration, over which the quantum system evolves: $\Delta T = \tau$.

Similar to the position and momentum operators introduced in Chapter 5, the energy operator—also known as the Hamiltonian, denoted by H—represents the energy observable of the system. Thus, if Henry wants to measure the *average* energy of his battery he should compute the following quantity:

$$< \psi \,|H|\, \psi >= \left(\sum_{i=0}^N < E_i|a_i\right) H \left(\sum_{j=0}^N a_j|E_j >\right) = \sum_{i,j=0}^N a_i a_j < E_i \,|H|\, E_j >$$

Here we used the fact that H, the energy operator, multiplies each energy state $|E>$ by its energy, and since each energy state is *orthogonal* to all other energy states, the calculation yields:

$$= \sum_{i,j=0}^N a_i a_j < E_i \,|E_j|\, E_j >= \sum_{i,j=0}^N a_i a_j E_j < E_i \,|E_j >= \sum_{i=0}^N p_i E_i = E.$$

Having introduced the *Hamiltonian*, we introduce the famous Schrödinger equation, which relates the dynamics (time change) of a quantum system to its energy:

$$H|\psi >= i\hbar \left(d|\psi >\right)/dt$$

On the left-hand side of the equation we have the *Hamiltonian*—the energy operator that operates on the quantum state. On its right-hand side we recognize $\hbar$, discussed in previous chapters, and $i = \sqrt{-1}$ from Chapter 5. On the right-hand side we have the derivative with respect to time $\frac{d}{dt}$ of the quantum state (also introduced in Chapter 5), which represents its *change in time*: how the quantum state evolves over an (infinitesimally small) time step dt.

To understand the dependence of energy on the evolution we must describe the probability amplitudes of the energy states in Henry's battery in terms of the Schrödinger equation:

$$H \sum_{j=0}^{N} a_j \,|\, E_j > \; = \; i\hbar \frac{d}{dt} \sum_{j=0}^{N} a_j \,|\, E_j >$$

$$\sum_{j=0}^{N} a_j H \,|\, E_j > \; = \; i\hbar \sum_{j=0}^{N} \frac{da_j}{dt} \,|\, E_j >$$

$$\sum_{j=0}^{N} a_j E_j \,|\, E_j > \; = \; \sum_{j=0}^{N} i\hbar \frac{da_j}{dt} \,|\, E_j >$$

$$a_j E_j = i\hbar \frac{da_j}{dt} \quad j = 0, \ldots, N$$

Thus, the Schrödinger equation reduces to a simple equation for each of the probability amplitudes: they change with time, proportionally to the amount of energy they represent. Without going into the details of how to solve this equation, we note that we need to integrate *over the derivative*, which results in $a_j(t) = e^{-\frac{iE_j t}{\hbar}}$.

According to these equations, each energy state evolves *independently of all the rest* (i.e., is *uncoupled* from all the rest). Yet, in the course of charging, the battery goes from the lowest energy state to higher ones. If we set the lowest energy to $E_0 = 0$, and assume the energy flow rate to be constant throughout the charging duration T, we obtain the following probability p_j to be in a state with energy E_j at *the end of the charging process* (the complete calculation will be given in subsequent chapters):

$$p_j = \frac{\sin^2 \left(\frac{E_j T}{2\hbar} \right)}{\left(E_j \right)^2}$$

This probability to find a certain energy deviation from the mean E is depicted in Figure 6.2.

The shape of this probability distribution has several important consequences. First, the energy state with the average (mean) energy E always has the highest probability. This is not surprising, since Henry charges his rocket with this amount of energy. Second, there is *non-zero probability for (almost) all other energy states*. Thus, even though Henry consumes energy E for the charging, there is

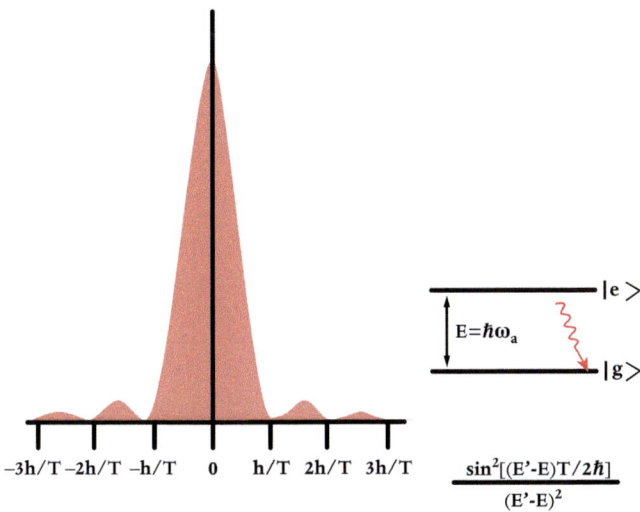

Fig. 6.2 The energy distribution described in the text as a function of time characterizes the transition from the excited $|e>$ to the ground $|g>$ state via the emission of a quantum: the energy distribution (called the sinc^2 function) is peaked at the resonant value $E = \hbar\omega_a$ (here set to be 0) but has a width and wiggles that vary inversely with $\hbar/T$, T being the time from the beginning of the process. The same (sinc^2 function) distribution characterizes the absorption of a quantum from the ground to the excited state, as in Henry Bar's quantum rocket that takes him from the ground floor to a high floor of the Pisa tower.

a chance that his rocket will receive either more or less energy than E. Third, the width of the central peak, which represents the energy uncertainty, shrinks with the duration of the charging T. In other words, when the charging duration is very long, the probability to obtain exactly the E state is very high. However, when the duration is extremely short there is a large probability of obtaining other energies.

The energy uncertainty is given by $\Delta E \geq \frac{h}{T}$, which is the width of the central peak. Thus, the *time–energy* uncertainty results in $\Delta E \Delta T \geq h$. It allows for an extraordinary effect, whereby a short duration of energy exchange with a total energy E results in a sizeable probability of finding states that have either more or less energy than E.

The short time interval in which Henry must charge his battery to help Alice results in high energy uncertainty: he can *jump either much higher or much lower than what is allowed by the energy he has invested in the charging*. Thus, our quantum superhero has exploited yet another unique and counterintuitive phenomenon of quantum physics.

PART II

Quantum entanglement and open quantum systems

Schred the Cat

The next day.

CHAPTER 7

What is Quantum Entanglement?

7.1 SCHRED AND HENRY ARE ENTANGLED

Until this adventure, Henry has been the only quantum superhero in our story. He has constituted a *single, simple* quantum system. Such a system exhibits the bizarre and wonderful behavior of a quantum superposition state that gives rise to phase interference, state collapse by measurement and uncertainty effects. Yet the world is made up of many systems, so that the question arises: what new and unique quantum phenomena are expected when two or more quantum systems interact? In the present adventure, two quantum characters interact. This happens because Henry has now constructed a new quantum suit—this time for his intelligent and curious cat, Schred, whom he has named after the famous quantum physicist, Schrödinger.

Equipped with its new quantum suit, Schred, with a little help from its human friend, splits into two quantum versions of itself and becomes the first quantum cat. But then, behaving according to its species' quirks, quantum Schred suddenly leaps onto Henry's lap. The crux of the situation is that only a single quantum version of Schred has leapt, while the other quantum Schred has remained still. Henry, wearing his quantum suit at that time, catches the leaping cat. Since only one of the two quantum Schreds has been caught by Henry, Henry then splits into two quantum versions of himself: the one that has caught the leaping Schred, and the one on which the (still) cat has not leapt. In our cartoons, the two couples are represented by different colors. One couple—the leaping Schred and the catching Henry—are colored red, while the other couple—the resting Schred and Henry—are colored blue. The joint state of Henry and Schred is called *a quantum entangled state*. Thus, the little incident that has played out between Henry and Schred has resulted in one of the most prominent phenomena in quantum mechanics: *quantum entanglement*. The nature of such phenomena has been a theme of heated debate, because of its conceptual importance and oddity. It is also a subject of enormous technological significance, as discussed further in this book.

Before delving into the mysterious world of entanglement, let us consider a simpler scenario involving multiple quantum systems; namely, *separable* quantum Schred and quantum Henry. Let us assume that both Henry and Schred have pressed their Split button, and no leaping has occurred. Schred would then be in a quantum superposition of, say, lying on the floor and pacing around, whereas Henry would also be in a superposition of his quantum versions; e.g., one sitting and one standing. Their joint system is now in a superposition of *four* possible states: *lying–sitting, lying–standing, pacing–sitting, pacing–standing*. This two-body system is composed of two *separate* single-body systems, with all possible combinations of their states present.

But what happens when one of the systems in this *separable* joint state is measured? If Schred is measured, then it will collapse to either a lying or a pacing cat. Let us assume that Schred collapses to the version lying on the floor. What does it tell us about Henry? The answer is 'nothing', for Henry is still in a superposition of his sitting and standing versions. This means that Henry is *decoupled* from Schred in this situation, hence any operation performed on Schred, such as measurement, does not influence Henry.

Let us compare this situation to our present adventure, in which the quantum cat has leapt onto his human friend. Their joint two-body system is now represented by only *two* possible states: *leaping–catching and resting–resting*, instead of the *four* possible states of a separable system. The difference between the two situations is that Henry and Schred are now *coupled*, so that their combined system cannot be described by the state of Schred alone and that of Henry alone, but rather by an *inseparable* state of Henry and Schred.

In the latter *entangled* situation, measurement has a completely different effect. Consider what has happened when Eve has measured Schred during its visit to her place. As in a single quantum- superposed system there are two equally probable, randomly determined, outcomes of this measurement; namely, Schred at Eve's place, or Schred with Henry. Let us assume, as in our story, that Eve's measurement has collapsed Schred to its (red) version with Henry. What does this measurement tell us about Henry? Since the red version of Schred is the one that has survived, whereas its blue version has disappeared, Henry can now only be in his red version. The blue version of their joint system has ceased to exist following the measurement-induced collapse, even though only Schred and not Henry has been measured. Remarkably, the fact that a blue Schred does not exist after this measurement means that the blue Henry has also disappeared. Both blue versions have vanished at once, so that Eve's measurement of Schred has resulted in Henry's collapse.

This is an illustration of one of the most extraordinary consequences of quantum entanglement, whereby measuring one of two entangled systems *immediately* collapses the state of the other system, even if the two are as far apart as we like, say, in different galaxies! *Quantum mechanics does not restrict the distance over which the collapse of an entangled state occurs for both systems at once.* Yet this does not mean that such collapse violates Einstein's causality principle, whereby *signaling is restricted by the speed of light and is not instantaneous.* Namely, one system does not acquire instantaneously any information from such collapse concerning the other system with which it has been entangled, since information does not travel faster than light. This issue will be elaborated in Chapter 14.

Let us disentangle the ingredients of the present intricate adventure from the start. Schred's new quantum suit has enabled it to become a quantum cat in a superposition of two states. One of these states leaps onto Henry, who then immediately splits into two quantum versions, thereby becoming entangled with Schred (the two versions are depicted in different colors). One of Schred's entangled versions runs towards Eve's apartment, whereupon Eve measures it. The measurement collapses the Schred–Henry joint-system state to the red version, thus manifesting the effect whereby measuring one system— here Schred—results in the immediate collapse of another remote system—here Henry.

One aspect of this plot remains unexplained. What has caused *Henry's* split in the first place? Or why does Schred's quantum leap affect Henry in such a peculiar way? In other words, *what causes the entanglement of quantum systems?* The answer to this important question is that entanglement occurs if two non-trivial requirements are satisfied. One requirement is that one system is affected *conditionally* on the state of the other system. To understand the meaning of this requirement, note that in our case Henry's state has been affected conditionally on Schred's leap. If Schred leaps, then Henry catches it; if Schred remains still, then Henry is unaffected. The other requirement is that one system is in a superposition—in our case, Schred. Thus, the fact that Schred both leaps *and* does not leap creates the situation in which Henry catches it *and* does not catch it at the same time. Schred's superposed behavior and Henry's conditional reaction results in Henry not only becoming superposed but also entangled with Schred.

Does entanglement happen often, or is it solely the experience of our quantum superheroes? Surprisingly, entanglement happens *all the time* all around us, at least on small (nanoscopic) scales. Whenever a superposed particle, such as a split electron, interacts with another system in a position-dependent way, such as with another electron, the two become entangled. This effect is so ubiquitous that quantum physicists in their laboratories try very hard to avoid it. Their main goal is to keep their precarious quantum systems, be it electrons, atoms or molecules, from becoming entangled in an *uncontrolled* fashion to other systems, such as electromagnetic radiation (photons) or colliding gas particles.

How crucial is it to keep quantum systems free of unwarranted entanglement with other systems? And what will happen to poor Schred? The answers to these questions will take up most of Chapters 8–11.

Entanglement Doom

If we exchange a furtive glance
As we rush on our way,
Remember that there is a chance
We'll never break away!
For quantum physics says, alas,
That our states get bungled
If we ignore that both of us
Forever stay entangled.

7.2 ENTANGLEMENT AND QUANTUMNESS

Von Neumann, who formulated the mathematical foundations of quantum mechanics (QM) between 1929 and 1932, just a few years after its inception, noted the peculiarity of quantum measurements (partly discussed in Chapter 5): If both the measured object and the measuring device are treated quantum-mechanically, then the states of the two, which may be assumed independent prior to the measurement, become inseparable thereafter; that is, they can only be described by a *joint state* and not by their individual states. The reason is that, in general, once we choose the kind of states measured by the device (such as momentum states), then each of the measuring-device states must become correlated to an eigenstate of the measured (object) observable, say, its spin orientation in the Stern–Gerlach experiment (Chapter 5). This implies that by observing (reading out) the measuring device, the object "collapses" (or is projected) onto the corresponding eigenstate. The oddity of quantum measurements culminating in Von Neumann's projection postulate (Chapter 4) may thus be reduced to that of the device-object quantum correlations. Von Neumann and his friend Wigner pondered deeply on the philosophical issues raised by such correlations (Section 7.3), but these issues came to the fore only after the appearance of the 1935 paper by Schrödinger in which he termed such correlations *entanglement (Verschränkung)*—a catchy name for a property that has since then been recognized to be one of the hallmarks of QM, on a par with the superposition principle and quantum coherence (Chapters 2–4).

Schrödinger turned towards the issue of entanglement in 1935 because of the last-ditch effort made that year by Einstein, together with his research assistant Nathan Rosen and the philosopher Boris Podolski (alias EPR), to shatter QM by arguing that this theory is *consistent but incomplete*. To prove their assertion, EPR quantum-mechanically analyzed two particles that are first in contact with each other but then separate over a large distance on account of their *free* counterpropagation. If we now measure the position of one particle we only know for certain the position of the other, but if we measure the momentum of one we only know with certainty the momentum of the other. Hence, the observable measured on one particle determines which observable of the other particle is known precisely, without uncertainty, and which is not (Figure 7.1). Yet, in the EPR parlance, such observables are the *elements of reality* of each particle, which should not change as a result of an action (e.g. a measurement) performed on another, distant, particle. They concluded that QM is "incomplete" because it does not specify *all elements of reality* of both particles!

The EPR paper (whose implications for information sharing between entangled objects will be discussed in Chapter 14) produced Schrödinger's response within a matter of months. Contrary to the EPR, Schrödinger stated that the "influence" of a measurement of one particle on another is to be expected regardless of their distance because, following their initial contact, they have become "entangled"; namely, their quantum states are correlated and inseparable.

In order to illustrate the bizarre nature of entanglement, Schrödinger added a paragraph at the end of his paper, outlining a challenging but principally allowed

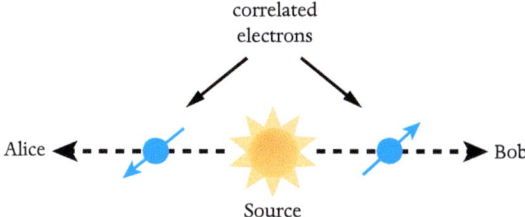

Fig. 7.1 A variant of the Einstein–Podolski–Rosen (EPR) suggested setup where the left and the right particles are entangled in their spin as they recede from each other (towards Alice and Bob, respectively). By measuring the z-component of Alice's spin, we can predict Bob's z- (but not x-) spin component.

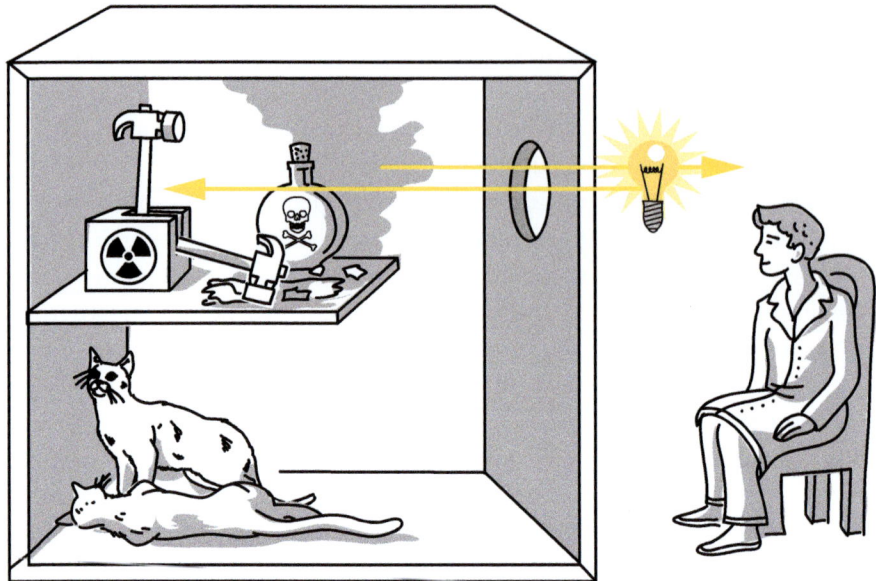

Fig. 7.2 Schrödinger's suggested setup wherein the live and dead states of a cat are entangled with those of a decayed or undecayed radioactive atom, respectively.

scenario wherein a radioactive atom may be entangled with a cat (Figure 7.2). The radioactive decay of the atom would trigger poison release that would kill the cat, but if we have not measured whether the atom has decayed or not, its states would be entangled with those of the cat. The two objects would be in a *coherent superposition* of two states: 1) a state of an undecayed atom and a live cat, and 2) a state of a decayed atom and a dead cat.

This outrageous scenario, which became known as "Schrödinger's cat", was meant to point out a central difficulty: if QM encompasses all reality, what prevents humans or animals from being in quantum superposition states in daily reality? We shall discuss possible responses to this question in subsequent chapters.

Yet even at this stage we may refer to the experimental possibility of fairly large and complex objects being entangled: The most interesting from the point of view of applications are *quantum computers*, whose advent is being announced more and more often. These are devices in which entanglement between many two-level particles (spins, see Chapter 5) is exploited for the performance of *coherently superposed computations in parallel*. Entanglement gives rise to an enormous number of such superposed states, and the projection of each state onto a chosen set of states (basis) of a detecting device represents a computation result, so that the number of results generated by the quantum computer per computation step can be astronomically larger than that of a conventional computer comprised of the same number of particles (provided this number is on the order of 100 or more). The predicted (albeit challenging) realization of quantum computers (Chapter 15) will extend entanglement from the sphere of conceptual curiosities to that of revolutionary technological applications and inevitably confront us with the peculiarity of quantum logic, whose philosophical implications were raised in Section 3.3.

7.3 ENTANGLED WORLD

As we saw in Chapter 1, quantum effects occur when an object changes its action by one or few units of $\hbar$. Since action change is the product of energy and time changes, even tiny energy exchange via interactions between objects can cause their entanglement, provided the exchange lasts long enough. Thus, any closed system of weakly interacting particles, no matter how large, may be considered to be fully entangled; i.e., be described by an incredibly complex many-body correlated state. This description goes beyond that of statistical physics of dilute systems, which commonly disregards such correlations in the system, replacing it by a "typical" single particle subject to repeated, randomly occurring interactions with other particles. This simple picture implies that macroscopic variables, such as the combined energy of two ensembles separated by a partition (wall), is the same as the sum of their individual energies after we remove the partition; i.e., is additive. The same additive property is supposedly true for their combined entropy measured by the number of states of the two ensembles.

Yet this picture is merely a convenient idealization which is being increasingly transgressed as our experimental and calculational capabilities improve. As a result, we can measure and analyze non-additive effects of two statistical ensembles concerning their energy or entropy. Admittedly, these are classical correlation effects, but their more subtle quantum counterparts based on entanglement are also showing up in current advanced experiments. E. Polzik (Denmark), M. Oberthaler (Germany) and N. Gisin (Switzerland) have all recently observed signatures of large numbers (billions, in the case of Gisin) of entangled atoms. Quantum computers (Chapter 15) that are now being developed would require thousands of quantum objects prepared in any desired entangled state—a highly challenging but not hopeless undertaking.

These considerations mainly concern entanglement on microscopic scales, but do they hold on much larger scales? There are now meticulously constructed experiments wherein entangled beams of light transfer their entanglement to hitherto uncorrelated detectors, which may be separated by hundreds or even thousands of miles (Chapter 14). Yet such sophisticated experiments are required because their goal is to fully control entanglement. If we keep in mind the photons exchanged between cosmic bodies, we may conclude that even distant stars are entangled with each other, although their entanglement is uncontrolled and may be masked by overwhelmingly larger classical effects.

In fact, this conclusion is supported by an enigmatic aspect of entanglement: it can persist even if there is no longer any interaction between the entangled objects and they are far apart. Einstein, Podolski and Rosen (EPR) faulted QM for this bizarre feature of entanglement which implies non-locality—some kind of synchronization or "collusion" between quantum objects regardless of their distance (Chapter 14). Yet, notwithstanding the EPR objections, non-locality has been experimentally verified, and so has one of its fascinating implications: quantum teleportation (Chapter 14). The persistence of entanglement between remote objects raises the question of whether there is a common origin to entanglement across the universe. Cosmology suggests such an origin: it tells us that the entire universe sprang out of an infinitely dense "singular" point (without spacetime dimensions) forming a quantum state of a unified field. Georges Lemaître (Belgium)—the originator of the Big Bang theory in the 1930s— referred to it as the "primordial atom". In the current version of this theory, spatial correlations between chunks of matter in the early universe (observed by satellite COBE) can be traced to quantum fluctuations in that primordial entity. It is not far fetched to presume that these fluctuations have generated an intricate web of persistent quantum entanglement across the universe shared by all objects, from elementary particles to astronomical bodies. The universe may well be a single, entangled—that is, inseparable—entity! It is at present impossible to find evidence for this unified view of the universe, because quantum corre- lations of the primordial universe are exceedingly feebler and far more fragile (Chapter 9) than their classical counterparts. Yet, in the future, perhaps with the advent of quantum computers (Chapter 15), we may be able to analyze the QM structure of such highly complex systems and develop tools for probing this structure despite its fragility. The existence of such hidden cosmic entanglement may then be revealed.

Well before the advent of QM, eighteenth-century scientists such as P. S. Laplace (France) regarded the universe as a single correlated entity—a giant machine whose countless constituents execute synchronized complex motions ruled by Newtonian mechanics. Would a quantum-correlated universe be prin- cipally different? It would indeed, according to the view advanced by D. Bohm whereby the world possesses a unique structure: it is a *quantum hologram* where each small (microscopic) piece encodes information on the world entire. This idea of Bohm concerning "implicit quantum order" has resurfaced in recent models of the universe. Clearly, this order hinges on the notion that the universe is entangled. If so, there may come a day when the quantum hologram of the world—the "state of everything"—unfolds in its full glory, revealing the minutest details of correlations on all possible scales of the universe in all its (covert and overt) dimensions.

But can such a state exist? This question is subtle, because the "existence" of wavefunctions is arguable. Mach and the other Vienna Circle Positivists would have been totally opposed to their existence had they lived to see Schrödinger's theory expounded, because a wavefunction is not directly measurable, and measurability was the sole criterion for physical relevance in the eyes of Positivists. Instead, a wavefunction encodes all information allowed by QM on a given object, and, by the same token, the "state of everything" would encode all information about the world. But who would possess this information? As argued in Section 8.3, the resolution of this question requires the acceptance of free will for observers of the world.

The description of the universe as a "quantum hologram" resonates with the ancient Hindu view of the "soul of the world" or the common essence of all things: the Brahman, which can be revealed at every level of the natural hierarchy. This Hindu view has found its way into Western philosophy, particularly into A. Schopenhauer's *The World as Will and Idea*. This view is part of a broader statement that lends itself to an analogy with quantum physics: *the identity of the Brahman with the Atman, the true human essence.* In the language of modern physics we may rephrase this statement as follows: *the quantum state (wavefunction) of the world is our innermost self.* Many would object to such a statement, but not those who regard our notions of the physical world as reflections of our "state of mind", as discussed in Chapter 8.

7.4 APPENDIX: ENTANGLING OPERATORS

A joint state of two systems can be described by either the bra-ket formalism or the matrix formalism (Section 4.4). We will discuss the operators that create entanglement in both formalisms, starting with the bra-ket.

Schred's quantum state, after its splitting, can be described by $| \, Schred >= \frac{1}{\sqrt{2}} (|L > + \, |R >)$, where $|L>$ represents the leaping Schred and $|R>$ represents the resting Schred. However, when composite systems are considered it is customary to denote each system by a subscript; e.g., $|Schred>= \frac{1}{\sqrt{2}} (|L>_s + |R>_s)$.

To account for states of two degrees-of-freedom—here Schred and Henry—we introduce the new symbol $\otimes$, which denotes the multiplicative product of the states of the two degrees of freedom. In this notation the initial product state, following Schred's split, is given by:

$$| \psi >=| Schred > \otimes | Henry >$$

In this case, $|Henry>=| S>_H$ is the sitting Henry, where the subscript represents Henry's state. We can now expand their joint state as follows:

$$| \psi >= \frac{1}{\sqrt{2}} \left(|L>_s \otimes |S>_H + |R>_s \otimes |S>_H \right)$$

Using the $\otimes$ operator may be redundant, however, because the system-specific subscript suffices for a more compact notation, resulting in the state:

$$| \psi >= \frac{1}{\sqrt{2}} \left(|L>_s|S>_H + |R>_s|S>_H \right)$$

Now consider the change in the state after Schred has leapt onto Henry, who then splits. It is tempting to describe Henry's new state as: $| Henry >= \frac{1}{\sqrt{2}}$ $(|C>_H + |S>_H)$, where $|C>$ denotes the catching Henry, but this is a wrong description, since Henry cannot be described separately from Schred. They are entangled, and their combined system requires a full description of both. Thus, after Henry has caught the leaping Schred their combined state becomes:

$$| \varphi >= \frac{1}{\sqrt{2}} \left(|L>_s|C>_H + |R>_s|S>_H \right)$$

In this new state, one of the sitting-Henry versions has caught the leaping Schred, while his other version remains sitting, corresponding to the version of Schred that has remained at rest.

Let us now consider a measurement described by the projection operator. In our story, Eve measures Schred's position, which then collapses to the red (resting) version. This measurement is represented by the projection operator $|R>_{ss}< R|$.

Let us consider what would happen if Schred did not leap onto Henry. Then this projection operator acting on the corresponding state would yield:

$$| R>_{ss}<R | \psi >= \frac{1}{\sqrt{2}} \left(|R>_{ss}<R|L>_s|S>_H + |R>_{ss}<R|R>_s|S>_H \right)$$

$$= \frac{1}{\sqrt{2}} | R>_s \left(0|S>_H + 1|S>_H \right) = \frac{1}{\sqrt{2}} | R>_s | S>_H$$

In this case, Henry would remain sitting. If Eve were to measure the other version of Schred, the corresponding projection operator would be $|L>_s<R|$, and Henry's state after the measurement would become:

$$| L>_{ss}<L | \psi >= \frac{1}{\sqrt{2}} \left(|L>_{ss}<L|L>_s|S>_H + |L>_{ss}<L|R>_s|S>_H \right)$$

$$= \frac{1}{\sqrt{2}} | L>_s \left(1|S>_H + 0|S>_H \right) = \frac{1}{\sqrt{2}} | L>_s | S>_H$$

In other words, the result of measuring Schred would not influence Henry's state, since their states are separable.

Let us consider the same scenario, but with Schred and Henry entangled. The red (resting) projection operator measuring Schred now yields:

$$|R>_{SS}<R \mid \varphi> = \frac{1}{\sqrt{2}}(|R>_{SS}<R|L>_S|C>_H + |R>_{SS}<R|R>_S|S>_H)$$

$$= \frac{1}{\sqrt{2}}|R>_S(0|C>_H + 1|S>_H) = \frac{1}{\sqrt{2}}|R>_S \mid S>_H,$$

while the leaping projection of Schred results in:

$$|L>_{SS}<L \mid \varphi> = \frac{1}{\sqrt{2}}(|L>_{SS}<L|L>_S|C>_H + |L>_{SS}<L|R>_S|S>_H)$$

$$= \frac{1}{\sqrt{2}}|L>_S(1|C>_H + 1|S>_H) = \frac{1}{\sqrt{2}}|L>_S \mid C>_H$$

Thus, Henry's state now depends on Schred's measurement result. This is very counterintuitive, since Henry is neither measured nor influenced. Eve and Schred are *elsewhere* still, and Eve's measurement of Schred influences Henry's state.

Now let us shift to a matrix notation that will help us understand the creation of entanglement. First, we must describe the entire Schred–Henry Hilbert space; i.e., the space spanned by all the possibilities of their composite two-body system. These states are $|R>_S|C>_H, |R>_S|S>_H, |L>_S|S>_H, |L>_S|C>_H$.

We describe this state by a 4-vector: $\begin{pmatrix} |R>_S|C>_H \\ |R>_S|S>_H \\ |L>_S|S>_H \\ |L>_S|C>_H \end{pmatrix}$. Thus, we have $\mid \psi> =$

$\frac{1}{\sqrt{2}}\begin{pmatrix} 0 \\ 1 \\ 1 \\ 0 \end{pmatrix}$ and $\mid \varphi> = \frac{1}{\sqrt{2}}\begin{pmatrix} 0 \\ 1 \\ 0 \\ 1 \end{pmatrix}$. The entanglement operator can therefore be

represented by a 4x4 matrix of the following form that yields $\mid \phi>$ when acting on $\mid \psi>$:

$$\begin{pmatrix} 1 & 0 & 0 & 0 \\ 0 & 1 & 0 & 0 \\ 0 & 0 & 0 & 1 \\ 0 & 0 & 1 & 0 \end{pmatrix} \frac{1}{\sqrt{2}} \begin{pmatrix} 0 \\ 1 \\ 1 \\ 0 \end{pmatrix} = \frac{1}{\sqrt{2}} \begin{pmatrix} 0 \\ 1 \\ 0 \\ 1 \end{pmatrix}$$

In this operator, the state of one system depends on the state of the other. Since one is in a superposition state, the other also "splits", but in a peculiar way that renders the systems entangled.

Schred and Henry Disentangle

CHAPTER 8

Entanglement, Decoherence and Which-Path Information

8.1 DECOHERENCE: THE DARK SIDE OF ENTANGLEMENT

In this and the following chapters we focus on the fragility of quantum coherence, which is the key to Henry's superpowers. First, we shall attempt to reveal the origin of this fragility: the ubiquitous tendency of quantum coherence to become corrupted and disappear by a mechanism termed *decoherence*.

To understand the origin of decoherence, let us examine Henry's attempt to sneak into Eve's residence, where his misunderstanding of the implications of quantum entanglement puts him in peril. Henry assumes, with misguided confidence, that he could, as in previous encounters with Eve, take advantage of his quantum coherence to shape the path so as to evade her detection, by inducing constructive interference along the selected path and destructive interference along the other path. However, in the previous encounters only Henry was in a superposition state; he was an *isolated* quantum object that interfered with itself. In more technical terms, both of Henry's versions were *indistinguishable*. The fact that they represent exactly the same quantum Henry and differ only in their relative phases enabled the interference. The extreme opposite situation is that of Henry and Schred, who cannot *interfere* with one another, being completely different objects.

If Henry succeeded in interfering before, what has gone wrong in the current scenario? The answer is that Henry and Schred have become entangled, i.e. inseparable. As discussed in Chapter 7, their entanglement precludes their independent behavior, dictating instead a behavior described by their *joint* quantum state. In our depiction of Henry's adventures, this behavior is denoted by the colors of the two objects.

Entanglement now reveals its ominous face. Since Henry is entangled with Schred, and Schred's versions are in two different places—Eve's and Henry's houses—Henry's two versions are now *distinguishable*. Even if both versions meet at the same place, one version (depicted blue) and the other (depicted red) are distinct, and hence Henry *can no longer interfere with himself*. In other words, because blue Henry is correlated to blue Schred (that is, at Eve's house) and red Henry is correlated to red Schred (that is, at Henry's house), when both Henrys meet, one can always tell them apart: the two Henrys are differently tagged by different Schreds. This *which-path information or tagging* causes the two Henrys to behave as *different quantum systems* that cannot interfere. As long as Henry and Schred are entangled, each of them loses *the individual quantumness*, conditioned by quantum coherence—an effect termed *quantum decoherence. Each of them* becomes, to all intents and purposes, a *classical* object.

Let us examine their quantumness loss more closely. Schred split into two coherently superposed versions (states) by a purely quantum effect. One version then jumped onto Henry and became entangled with him, again by a quantum effect. Now Henry's superposed states can no longer interfere, which means they no longer have quantum coherence. Thus, by becoming entangled to Schred, our superhero has lost his quantum powers.

However, all is not lost, for Schred, being a very smart cat, understands that its human friend is in distress and comes to his rescue. By reuniting its two (red and blue) versions, Schred has *disentangled* itself from Henry. Their joint state, after red Schred's jump into Eve's house, has consisted of two superposed terms: (blue Henry–Schred at Eve's house) and (red Henry–Schred near Henry). Then, after red Schred has reunited with blue Schred at Eve's house—i.e., they have joined at the same location and recombined—the joint state becomes a product of two independent states: a superposition of blue and red Henrys and a complete Schred at Eve's house.

This disentanglement from Schred has made it possible for Henry to interfere just in time to evade Eve's guard and recover the precious briefcase Eve has stolen from his friends.

While grooming itself at Eve's, Schred accidently presses the Split button on its quantum cat-suit. Suddenly there are several blue Schreds. Can this blue Schred's splitting influence Henry in any way? As we have shown, Schred's recombining with itself frees Henry and Schred from their entanglement. But what would happen if one of Schred's blue versions got away? Henry and Schred would still be entangled, since not all of Schred's versions recombined. The lost blue version of Schred would make the two versions of Henry distinguishable, therefore void of quantum powers. Fortunately, Schred, being a sensible cat, has recombined all its versions, thus restoring Henry's quantumness.

Henry's and Schred's adventures have abundant analogs in our world. Consider an atom whose state has been split by the methods of modern physics (Chapter 2) into a superposition of its ground and excited electronic-energy states. To prove that it is a quantum superposition state, an experiment is performed that shows how constructive and destructive interference are controlled via a change in the relative phase of the superposed states (Chapter 3). The experiment is repeated many times to ascertain its result. What the experimenters are unaware of is that between two consecutive experiments a stray atom has collided with the atom at hand, and the two have become entangled. To make matters worse, the stray atom has since gone its way. The second interference experiment then fails, as the examined atom no longer exhibits its interference patterns, since its two states *cannot interfere* as long as they are entangled to the states of the stray atom.

In order to restore coherence in the examined atom, the experimenters may attempt, like Schred, to unite the two quantum states of the stray atom, for example, by making these two quantum states identical ("degenerate"). Alas, such attempts would be to no avail, for the stray atom cannot be recovered. The quantum superposition state of the examined atom has decohered; i.e., it has irretrievably lost its coherence due to its entanglement with the stray atom that is now beyond the experimenters' control and can be viewed as part of the *environment*.

This *induced decoherence by the environment* is the biggest hurdle en route to the crafting of increasingly larger and more complex quantum systems, as required, for example, for quantum computers (Chapter 15). Any stray atom or photon from the environment may become entangled with the quantum system of interest and then be lost somewhere, carrying away information that distinguishes between the states of the quantum system (or, equivalently, the paths they have taken) and thus render this system classical, to all intents and purposes.

Is there no hope of maintaining quantum coherence in complex systems? One obvious and common way to avoid decoherence, albeit at great effort and cost, is to isolate the quantum system from the environment. For atoms this is done by placing them in vacuum chambers, or even better, in magneto-optical traps, so as to reduce their chances of colliding with stray atoms. Faraday cages may be used to shield the system from electromagnetic fluctuations that also cause decoherence effects (Chapter 9). Electron spins confined to nanometric-size nitrogen vacancies in a diamond are currently the quantum systems that are best isolated from environmental decoherence, so that their coherent superpositions can live for minutes if not hours. It is also advantageous to prepare superposition states of photons that are robust because their interactions with the environment are extremely weak, and thus they maintain their coherence over long distances and long durations.

In addition, there are clever ways to overcome decoherence by exploiting quantum effects, as described in Chapters 10–12. But before Henry can harness the power of such effects to fight decoherence, he must overcome yet another obstacle that Eve and the environment confront him with in Chapter 9. We shall discuss how to overcome these difficulties and thereby allow physicists to push the scale on which quantum coherence is manifest towards increasingly larger and more complex objects. The prospects of attaining a man-size scale of quantum coherence, such as displayed by Henry, are, however, very slim in view of the technological obstacles they present which at the moment are insurmountable.

8.2 DECOHERENCE AS WHICH–PATH DISTINGUISHABILITY AND ENTANGLEMENT WITH THE ENVIRONMENT

The entanglement with Schred has tampered with the phases of Henry Bar's superposition state. Similar processes whereby the phases of superposition states are corrupted occur naturally and unavoidably in any quantum system of interest, such as a qubit. If we follow its state in time, then sooner or later its interaction with its surrounding (environment) will cause the loss of its quantum phase coherence, termed *decoherence*.

The notion of decoherence was introduced by J. Von Neumann in his groundbreaking book on the foundations of QM in 1932. This notion followed from his discussion of quantum measurement effects (Chapter 5): an ideal measurement projects a quantum superposition state onto an eigenstate, thereby destroying the phase coherence that existed among the eigenstates in the superposition. This process constitutes decoherence in its strongest sense. It occurs since the system and the measuring apparatus become entangled for the measurement to take place and then the apparatus is ignored, rendering the quantum state of the system mixed, or impure, without phase coherence.

Subsequent work has yielded further insights into decoherence. Firstly, it has been shown (starting with D. Bohm, then in the USA in the late 1940s) that this process does not have to be "premeditated", as the environment with which the system interacts can play essentially the same role as a measuring apparatus. The results of "measurements" by the environment on the system are not accessible to us, but their effect is to bring about purity or coherence loss by the state of the system, just as a measuring apparatus would. Since every quantum system is in contact with some environment, and the vulnerability of quantum states to decoherence typically grows with the complexity of the system (see Chapter 15), decoherence has come to be viewed as the main reason for the lack of quantumness in nearly all natural phenomena that we encounter, as proposed by W. Zurek (USA) in the 1990s.

As a rule, the early works in QM which we surveyed in Chapters 2–5 did not discuss decoherence, with the exception of the Einstein–Bohr controversy as to whether a tiny kick imparted by an electron to one of two slits (in a plate mounted on wheels) would cause a kick-back of the slit that is large enough to wash out the interference pattern created by many such electrons on a screen. Bohr and Rosenfeld calculated that whenever the traversed slit is clearly disclosed by a shift of the plate caused by the electron, its kick-back exerts a force strong enough to appreciably change the direction of the electron to the extent that the interference pattern is modified. Since the size and the direction of this force unpredictably change from one electron to another, the force is effectively random. For decades thereafter, decoherence was identified with the effect of *random force* on the motion of a quantum wavepacket, causing the destruction of its phase-coherent character (Figures 5.1 and 8.1).

Yet many years later, in the 1980s, the identification of decoherence with random-force or kick effects was refuted by M. O. Scully (USA) with his collaborators H. Walther, K. Druehl and B. Englert (Germany). They proposed a "which-path" measurement of a quantum object—say, a flying atom—that could leave a trace of its whereabouts by emitting a photon along the left-hand or

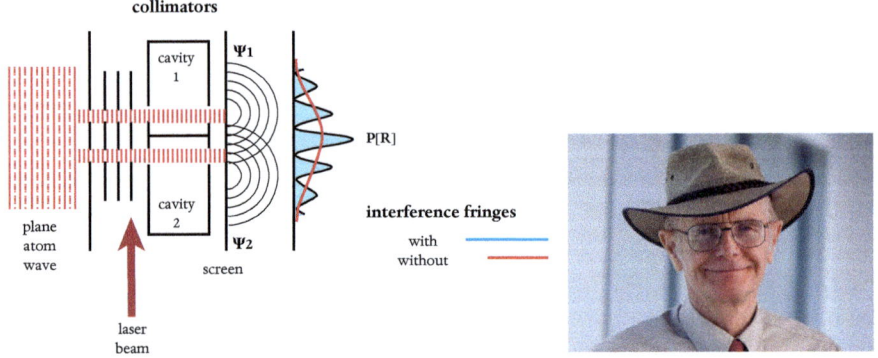

Fig. 8.1 The quantum eraser setup of Scully (right) et al. (see text).

right-hand path in a two-path setup (Figure 8.1). In their setup, an excited atom is launched in a state with well-defined momentum, so that according to the Heisenberg uncertainty relation it is completely spread out in space (Chapter 5). Two slits symmetrically located in the direction transverse to its propagation axis allow the atomic wavepacket to pass through either cavity 1 or cavity 2 (the left-hand or right-hand path, respectively). QM predicts that the atom is then in an equal, coherent, superposition of states (waves) passing through cavity 1 or 2. The two waves recombine as the atom emerges from the cavities, but before its emergence the atom is likely to emit a photon in one of the cavities. It is easy to impose conditions such that the kick-back causing momentum change of the recoiling atom by the emitted photon be negligibly small compared to its original momentum. Then, there is no appreciable random force that might blur the interference fringes which many such atoms leave on the screen. And yet the fringes are completely wiped out in this case, according to QM, even though the content of the cavities is not examined! The disappearance of the interference pattern may then be ascribed to information deposited in the measuring apparatus (here, by a photon in one of the cavities) rather than to a physical influence of the apparatus on the object (here, the atom).

The information recorded by the apparatus thus bears witness to the path traversed by a quantum object in the abstract space of its possible states or configurations ("configuration space") or, in other words, tags the actual path taken by the object. This information recording or tagging was identified by W. Zurek as the essence of decoherence. Zurek's simplest model was that of a spin-½ particle whose energy states are aligned with (spin-up) or against (spin-down) a magnetic field along the z-axis, respectively. This spin-½ object interacts with another spin-½ particle that serves as a meter (measuring apparatus). Suppose the object, prior to the interaction, is in a coherent superposition of spin-up and spin-down states. Following an interaction that attracts counteraligned spins, the two particles become entangled. The spin-up and spin-down states of the object correlate to the opposite states of the meter, so that the observation of a particular state of the meter discloses or tags the state of the object. Remarkably, even without reading out the meter state, the mere availability of information on the states of the object destroys their superposition.

Zurek applied the same approach to the effect of an environment on a quantum object: the environment in this approach is an uncontrolled apparatus (meter) that incessantly watches and measures the object. The key notion in this approach is that the environment determines the "pointer basis". In the example above it is the alignment or counteralignment of spins with the z-axis, along which the object and the meter spins become correlated. The fact that a superposition of spin-up and spin-down states of the object in this basis is destroyed (decohered) by the interaction with the environment (meter) implies that only the spin-up or spin-down states along z are robust against environment effects.

This robustness selects or singles out the preferred pointer basis in the presence of an environment. This *environment-induced selection* of a robust basis has been abbreviated as *einselection* by Zurek. It is commonly viewed as a resolution of the Schrödinger-cat paradox: superpositions of distinct states of a macroscopic object undergo extremely rapid decoherence via their interaction with the environment, hence they are never observed in daily life. However, as will be shown in Chapter 12, there are means of effectively countering or controlling decoherence and thereby pushing the observability of superposition states towards the realm of increasingly larger and more complex quantum objects.

8.3 ON INFORMATION AND FREE WILL: DO WE LIVE IN A QUANTUM MATRIX?

Paradoxically, entanglement is not only the hallmark of quantumness in a composite system, but is also the key to the *demise of quantumness*, alias *decoherence*, in each of the entangled objects. In such a system, one of the objects can play the role of an observing or measuring device (henceforth called a meter), and the other that of the observed object. Since each state of the meter is separately correlated with a corresponding eigenstate of the object, a coherent superposition of the object eigenstates would be destroyed—or in other words, decohered—by this entanglement with the meter. As a result, if we confine our attention to the meter, disregarding the object, we detect one of its eigenstates in each measurement unpredictably—i.e. randomly—apart from the known probability of the occurrence. Correspondingly, the same is true for the observed object.

Alternatively, decoherence can be viewed as the tagging of alternative evolution paths that are (at some point of time) superposed in a quantum state of an object. This tagging arises from entangling the object with a meter. Simply ignoring (averaging over) the states of the meter decoheres the object's state, turning it into an incoherent (or, equivalently, random-phased) mixture of the alternative paths. Contemporary approaches to QM regard decoherence as a transformation of *quantum information* stored in a coherent-superposition state of the object (a qubit, in the simple case of two superposed energy states) into *classical information* encoded in the probabilities of the individual states that formed this superposition.

Here we would like to go further by posing the disconcerting question: *who possesses this (classical or quantum) information on observed/measured objects?* If a distinction is drawn between the "world of information"—commonly identified with the cyberspace or the virtual world—and the "real" (material?) world, then the question is: who partakes in the virtual world? An obvious answer is: computer users—and even more so, computer experts and scientists that generate the content of the virtual world. But since in QM the mere act of observation, even if its outcomes are ignored, changes the information, we must drastically broaden the scope of the virtual world to include all observers that receive, process and share information. Humans, animals, plants, bacteria— all life forms have the capacity (and the necessity) to manipulate and even create information by reducing the entropy (randomness or disorder) in their organisms through metabolism or reproduction.

To these information bearers and processors we must add the artificial extensions of human intelligence. Yet the deeper question is: are live or artificial observers able to create and change information at will? In terms of observation, do observers have the freedom to choose the subject of their observation and the measurement basis?

Clearly, these questions reflect the millennia-long controversy of *free will*. Do I decide on my actions (within the constraints of reality) or are my decisions illusory, whereas in truth my actions are predetermined by a certain (albeit unknown) pattern I am programmed to follow? Adherents of determinism in philosophy (notably, the seventeenth-century Dutch philosopher Spinoza, who viewed our individuality as an insignificant, predefined mode of a universal hierarchy) and physics (such as the eighteenth-century French scientist P. S. Laplace, who vouched to determine the positions and momenta of all bodies in the universe if their initial values are given) could never come to terms with their free-will adversaries (such as the German philosophers Kant and Schopenhauer, who stressed our individuality) or the proponents of indeterminism manifested by randomness in physics (notably, Boltzmann).

Yet QM, on the face of it, appears to allow for the coexistence of these two contradictory approaches: On the one hand, the quantum state of an independent object evolves deterministically, being governed by the Schrödinger equation, while on the other hand the interaction of one object with another that results in their entanglement brings about a random, unpredictable evolution of each of them.

Furthermore, the essence of Bohr's complementarity (Chapter 7) is that the choice of the observable measured by the meter determines the possible results of measurements of any chosen observable on the object, and each choice may entail completely different measurement outcomes. But is there truly a fundamental difference between the evolution prior and posterior to the entanglement of two objects, or, equivalently, between pre- and post-measurement evolution? If we adopt Von Neumann's standpoint, then the crucial moment is when the "natural" evolution is replaced by the observer's mental selection of one of the alternative results of the evolution. Obviously, Von Neumann and Bohr believed in free will or freedom of choice. By contrast, according to one of the staunchest adherents of Everett's many-world interpretation (Chapter 4), the physicist L. Vaidman (Israel), Everett's approach leads to complete determinism, by which the multiverse is described by one quantum state that includes all the observers and the observed, so that in a given world there is a unique, predetermined result of each action taken by each observer on all objects. The price of this determinism is the unimaginably large number of worlds involved in each such action.

The enormous complexity of Everett's description of the multiverse precludes putting its underlying determinism to a test. As shown by the US physicist S. Lloyd, even the test of free will on a conventional Turing machine (a computer) is too complex to be conclusive! The preference of free will to determinism (or vice versa) is then a matter of taste. Do we (past, present and future) inhabitants of the universe lead an involuntary existence as part of the "state of everything" that is ruled by a quantum matrix (Section 8.4)? Or, are we all (bacteria, plants, animals and humans) masters of the situation, acting autonomously by choosing the basis for our observations and acquiring information through such observations?

If we extend these considerations to the entire universe and treat the observer quantum-mechanically, as befits the universality of QM, then the observer's entanglement with the universe implies that the events that befall each observer—each living creature—must be random. It is an enticing idea that entanglement is the source of all randomness in life. But is it legitimate for an observer to consider the universe as an independent reality to be or not to be entangled with? Any discussion of this question brings us inevitably into the realm of philosophical issues—primarily, how can I perceive the world in spite of my being part of it? Another puzzling issue is whether all observers must agree on their observations or perception. If not, then the world is not objectively knowable.

In European philosophy since R. Descartes (France, seventeenth century), doubt has frequently been cast on whether the world is knowable. Yet no philosopher has gone so far as the eighteenth-century Irish Bishop G. Berkeley, who asserted that there is actually nothing to know about the world—there is no reality outside our consciousness. He formulated his assertion in Latin as *Essere est percipi*—"to be is to be perceived". Berkeley entrusted perception to divine providence as the only assurance of all things to exist. Otherwise, would a tree fall down if there is no one in the forest to observe it? A remarkably similar conclusion had been reached a millennium earlier by the Buddhist thinker Vasubendhu in India. His teaching grew out of the ancient Hindu view that our perception of reality is illusory (Section 7.3) and that there is no separation between the external world and human consciousness.

These ideas evoke the cryptic verse in Ecclesiastes (chapter 3, verse 11) that Bishop Berkeley knew so well: ". . . he hath set **the world in their heart**, so that no man can find out the work that God maketh from the beginning to the end". Does it mean that the world is in our minds (hearts) and that we have no other knowledge of it?

The anxiety that there is no reality outside of our consciousness is described in Lewis Carrol's *Alice Through the Looking Glass*, in which Alice is warned not to wake the Red King from his sleep. "You are only a thing inside his dream," she is told, and if he wakes "you'll go out with a bang like a candle."

But who is the Red King that dreams up our world? According to Von Neumann's interpretation of QM, there is no world without an observer, whose mind disentangles the observed quantum state from that of the observing apparatus. Hence, in this view, *the observer plays an active, essential role in keeping the world intact.*

Yet this view smacks of subjectivity, or even solipsism. Does it mean that the world ceases to exist when I, the observer, shut my eyes? And what can vouch for an agreement between the perceptions of (or the information possessed by) many observers, so that there is but one reality? Bishop Berkeley attributed this agreement among observers and the reality of the world to God's mind.

There is, however, another approach to the objectivity of human perception, by the Alexandrian philosopher Plotinus, the founder of the Neo-Platonic School, c.200 AD. Plotinus asserted that there is a *collective form of human thinking*, shared by all mankind—the idea of the existence of the world, a divine idea that had preceded creation. Under proper guidance we may strive to regain the pristine simplicity and truth of this idea, thereby ascending to the common origin of all human consciousness. Can we cast this philosophy in modern terms, whereby we, individual observers, share an *objective mode* of the perception of the world as expressed by its quantum state, while retaining our *subjective* freedom by choosing the observables of our interest? Meaningful science must be objective, but can we take its objectivity for granted? We leave the answer to the readers' discretion.

We are still puzzled by these questions, raised in the concluding verse.

Who observes the observer?

Big Brother watches all, they say,
And decoheres each quantum path
Of atoms soaking in a "bath"
That makes them err and lose their way.

But who's Big Brother in our world
Observing objects first to last?
It's the environment, we're told,
That sweeps away our quantum past.

Yet the environment must too
Abide by nature's quantum laws,
And therefore it's unclear: just who
Is our Big Brother? Anyone knows?

8.4 APPENDIX: COMPLEMENTARITY BETWEEN VISIBILITY AND DISTINGUISHABILITY

In this appendix we mathematically investigate the interference exhibited by two quantum systems as a function of their entanglement, which we treat as a continuous parameter that may range between 0 (no entanglement) and 1 (full entanglement).

Tracing out a system. The first essential step in this investigation is to explain a fundamental operation in quantum mechanics of multipartite systems; namely, *trace*. Let us consider a bipartite system composed of two qubits—the first spanned by states $|\downarrow\rangle_1, |\uparrow\rangle_1$, and the second by states $|g\rangle_2, |e\rangle_2$. We have already discussed what a joint state looks like, but now we write it as a *density matrix*:

$$
\rho^{(1+2)} =
\begin{pmatrix}
\rho_{\downarrow g,\downarrow g} & \rho_{\downarrow e,\downarrow g} & \rho_{\uparrow g,\downarrow g} & \rho_{\uparrow e,\downarrow g} \\
\rho_{\downarrow g,\downarrow e} & \rho_{\downarrow e,\downarrow e} & \rho_{\uparrow g,\downarrow e} & \rho_{\uparrow e,\downarrow e} \\
\rho_{\downarrow g,\uparrow g} & \rho_{\downarrow e,\uparrow g} & \rho_{\uparrow g,\uparrow g} & \rho_{\uparrow e,\uparrow g} \\
\rho_{\downarrow g,\uparrow e} & \rho_{\downarrow e,\uparrow e} & \rho_{\uparrow g,\uparrow e} & \rho_{\uparrow e,\uparrow e}
\end{pmatrix}
$$

Here, for example, $\rho_- \{\downarrow g, \uparrow e\} = |\downarrow\rangle |g\rangle \langle\uparrow|\langle e|$. This density matrix accounts for all the possible ways of occupying these pairwise combined states and thus for all information regarding the two-qubit state. Here we ask: what happens when we do not know anything about (because there is no access to) one qubit? The proper mathematical operation that corresponds to this situation is *tracing out* (or tracing over) the qubit we know nothing of. The trace operation is represented as the sum over the probabilities of the basis states of the traced-over qubit 2, resulting in the following state of qubit 1:

$$\rho^{(1)} = Tr_2\rho^{(1+2)} = \sum_{i=g,e} 2\langle i|\rho^{(1+2)}|i\rangle_2 = \langle g|\rho^{(1+2)}|g\rangle + \langle e|\rho^{(1+2)}|e\rangle$$

This operation is *non-unitary*—an incoherent sum of the *probabilities* of the states of qubit 2. Namely, as opposed to a quantum superposition of two possible states, where we do not know which one is actually realized in each run, tracing out treats our lack of knowledge about one qubit as *classical* uncertainty.

In pictorial form, the 2×2 density matrix that now represents qubit 1 is the sum of the "blue" and "red" 2×2 matrices:

Here the "red" 2×2 matrix includes only $|g\rangle\langle g|$ terms, and the "blue" 2×2 matrix includes only $|e\rangle\langle e|$ terms.

In the same manner, we may write the state of qubit 2 that is obtained upon tracing out qubit 1 as

$$\rho^{(2)} = Tr_1\rho^{(1+2)} = \langle\downarrow|\rho^{(1+2)}|\downarrow\rangle + \langle\uparrow|\rho^{(1+2)}|\uparrow\rangle$$

which is represented pictorially by the sum of the following "red" and "blue" 2×2 matrices:

No entanglement: perfect visibility of interference. Having introduced the *trace* operation, we may reconsider Henry's state prior to Schred's frivolous activities. To describe interference in a general way, let us denote Henry's state as being oriented up and down, both in bra-ket and matrix notations:

$$|\downarrow\rangle = \begin{pmatrix} 1 \\ 0 \end{pmatrix} \qquad |\uparrow\rangle = \begin{pmatrix} 0 \\ 1 \end{pmatrix}$$

where the splitting and recombining operators are, in matrix form:

$$S = \frac{1}{\sqrt{2}} \begin{pmatrix} 1 & -1 \\ 1 & 1 \end{pmatrix}, R = \frac{1}{\sqrt{2}} \begin{pmatrix} 1 & 1 \\ -1 & 1 \end{pmatrix}$$

A compact pictorial description of an interference setup, *à la* Stern–Gerlach (Chapter 3) is as follows:

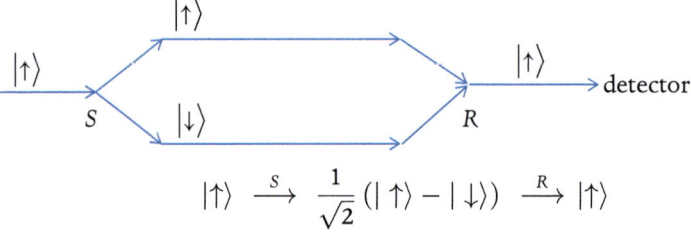

$$|\uparrow\rangle \xrightarrow{S} \frac{1}{\sqrt{2}} (|\uparrow\rangle - |\downarrow\rangle) \xrightarrow{R} |\uparrow\rangle$$

Here, the detector measures the probability to get $|\uparrow\rangle$, which in this case is equal to 1.

We next reintroduce the phase-dial operator (Chapter 3), but in a more general form:

$$|\uparrow\rangle \rightarrow e^{i\phi/2} |\uparrow\rangle , |\downarrow\rangle \rightarrow e^{-i\phi/2} |\downarrow\rangle$$
$$P = \begin{pmatrix} e^{-i\phi/2} & 0 \\ 0 & e^{i\phi/2} \end{pmatrix}$$

Here we have used the complex exponent notation, which is given by $e^{i\phi/2} = \cos \phi/2 + i \sin \phi/2$, where ϕ is the phase and $i = \sqrt{-1}$. In this notation we may follow the evolution of an initial spin-up state that traverses this setup, as depicted here:

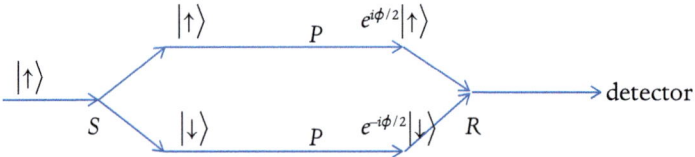

$$|\uparrow\rangle \xrightarrow{S} \frac{1}{\sqrt{2}}\left(|\uparrow\rangle - |\downarrow\rangle\right) \xrightarrow{P} \frac{1}{\sqrt{2}}\left(e^{i\phi/2}|\uparrow\rangle - e^{-i\phi/2}|\downarrow\rangle\right)$$

$$\xrightarrow{R} \cos\frac{\phi}{2}\,|\uparrow\rangle + i\sin\frac{\phi}{2}\,|\downarrow\rangle$$

Now the probability to detect $|\uparrow\rangle$ is given by the projection probability onto this state at the detector : $p\left(|\uparrow\rangle\langle\uparrow|, \phi\right) = \cos^2\frac{\phi}{2} = \frac{1+\cos\phi}{2}$.

This expression is known as the *Ramsey interference pattern*, which denotes the detection probability of spin-up as a function of the phase difference. The maximal contrast of this sinusoidal pattern, defined as the difference between the maximum (peak) and the minimum (trough) probability values, is called "visibility":

$$V = p\left(|\uparrow\rangle\langle\uparrow|, 0\right) - p\left(|\uparrow\rangle\langle\uparrow|, \pi\right) = 1 - 0 = 1$$

This case thus corresponds to *perfect visibility*.

In this case the two paths differ by a phase, but are otherwise equivalent. Namely, we cannot deduce from the interference pattern whether the spin-up particle arrived at the detector via the upper or the lower path: we may interchange the paths without changing the pattern. This setup is an interferometer with perfect visibility, because it does not record the actual path information!

Entanglement reduces visibility. We next investigate the consequences of our latest twist in the plot—quantum entanglement between two qubits—on the interference pattern. To this end, we let qubit 2 undergo a flip operation, denoted by WP (for reasons explained below), only along one path:

$$|g\rangle \rightarrow |e\rangle \, , \, |e\rangle \rightarrow |g\rangle$$

$$WP = \begin{pmatrix} 0 & 1 \\ 1 & 0 \end{pmatrix}$$

Thus, WP is a *path-conditional* operator. That is, its operation encodes which-path (WP) information: it *tags* each path of qubit 1 differently. The initial product state wherein qubit 1 is in the spin-up state and qubit 2 is in state g, is entangled by WP. The overall evolution of this state is then given in the previous setup by:

$$|\uparrow\rangle|g\rangle \xrightarrow{S} \frac{1}{\sqrt{2}}\left(|\uparrow\rangle|g\rangle - |\downarrow\rangle|g\rangle\right) \xrightarrow{WP} \frac{1}{\sqrt{2}}\left(|\uparrow\rangle|e\rangle - |\downarrow\rangle|g\rangle\right)$$
$$\xrightarrow{P} \frac{1}{\sqrt{2}}\left(e^{i\phi/2}|\uparrow\rangle|e\rangle - e^{-i\phi/2}|\downarrow\rangle|g\rangle\right) \xrightarrow{R} \frac{1}{2}\left(e^{-i\phi/2}\left[|\downarrow\rangle - |\uparrow\rangle\right]|g\rangle\right.$$
$$\left. + e^{i\phi/2}\left[|\downarrow\rangle + |\uparrow\rangle\right]|e\rangle\right)$$

The spin-up state of qubit 1 that is correlated to the g-state of qubit 2 is unequivocally associated with the lower path. The orthogonal spin-down state of qubit 1 that is correlated to the e-state of qubit 2 is likewise associated with the upper path. Because of their orthogonality, the states associated with the two paths do not interfere at the detector, hence no interference pattern is expected.

Since we only observe the interference of qubit 1, visibility is defined in the $|\downarrow\rangle, |\uparrow\rangle$ space and not for $|g\rangle, |e\rangle$. We thus need to "get rid of" qubit 2 in the entangled quantum state. This is done by the *trace* operator described above, which acts on the joint two-system state, which is denoted by ρ. This involves taking the *sum of the probabilities of* $|g\rangle, |e\rangle$, i.e. squaring the probability amplitude of $|g\rangle$ and adding to the square of the corresponding probability-amplitude of $|e\rangle$. It is easy to see that since we sum the probabilities, rather than the *probability- amplitudes*, all the terms involving the phase factors $e^{i\phi/2}$ drop out, since $\left|e^{i\phi/2}\right|^2 = 1$. This results in the following visibility:

$$\text{Detector}: p\left(|\downarrow\rangle\langle\downarrow|, \phi\right) = \left|\langle\downarrow|\text{Tr}_{eg}\rho|\downarrow\rangle\right|^2 = \frac{1}{2}$$
$$V = p\left(|\downarrow\rangle\langle\downarrow|, 0\right) - p\left(|\downarrow\rangle\langle\downarrow|, \pi\right) = \frac{1}{2} - \frac{1}{2} = 0$$

Namely, the entanglement erases completely the interference pattern, because of the orthogonality of the states associated with the two paths, as discussed above.

In order to measure the entanglement we define the *distinguishability* measure, which quantifies the effect of the which-path (WP) operator on the initial state:

$$D = \sqrt{1 - \left|\langle g|WP|g\rangle\right|^2} = 1$$

It is thus evident that maximal distinguishability corresponds to minimal visibility.

Is there a more general relation between these two measures? In order to answer this question we replace the former entangling flip operator by a more general form, in which g and e undergo the following transformation parameterized by a continuous parameter α (whose physical meaning will be discussed later):

$$|g\rangle \rightarrow \cos\alpha \, |g\rangle + i\sin\alpha \, |e\rangle$$
$$|e\rangle \rightarrow \cos\alpha \, |e\rangle + i\sin\alpha \, |g\rangle$$
$$WP = \begin{pmatrix} \cos\alpha & i\sin\alpha \\ i\sin\alpha & \cos\alpha \end{pmatrix}$$

Here, $\alpha = 0°$ means no entanglement and $\alpha = 90°$ means full entanglement. The evolution in this setup then acquires the following form:

$$|\uparrow\rangle|g\rangle \xrightarrow{S} \frac{1}{\sqrt{2}} (|\uparrow\rangle|g\rangle - |\downarrow\rangle|g\rangle)$$

$$\xrightarrow{WP} \frac{1}{\sqrt{2}} ((\cos\alpha|g\rangle + i\sin\alpha|e\rangle)\,|\uparrow\rangle|e\rangle - |\downarrow\rangle|g\rangle)$$

$$\xrightarrow{P} \frac{1}{\sqrt{2}} (e^{i\phi/2}|\uparrow\rangle (\cos\alpha|g\rangle + i\sin\alpha|e\rangle) - e^{-i\phi/2}|\downarrow\rangle|g\rangle)$$

$$\xrightarrow{R} \frac{1}{2} (|g\rangle [|\uparrow\rangle (e^{i\phi/2}\cos\alpha + e^{-i\phi/2}) + |\downarrow\rangle (e^{i\phi/2}\cos\alpha - e^{-i\phi/2})]$$

$$+ ie^{i\phi/2}\sin\alpha|e\rangle (|\uparrow\rangle + |\downarrow\rangle))$$

The detection probability and visibility of this composite state are, respectively:

$$\text{Detector}: p\,(|\downarrow\rangle\langle\downarrow|, \phi) = \frac{1}{4}\left|(e^{i\phi/2}\cos\alpha + e^{-i\phi/2})\right|^2 + \sin^2\alpha$$
$$= \frac{1 + \cos\alpha\cos\phi}{2}$$
$$V = p\,(|\downarrow\rangle\langle\downarrow|, 0) - p\,(|\downarrow\rangle\langle\downarrow|, \pi) = \cos\alpha$$

The corresponding distinguishability is:

$$D = \sqrt{1 - |\langle g|WP|g\rangle|^2} = \sin\alpha$$

The result is an amazingly simple relation between visibility (the measure of interference or quantum coherence between alternative paths) and distinguishability (the measure of entanglement):

$$D^2 + V^2 = 1$$

This means that the more entanglement the system has with the "environment", here represented by another qubit, the less pronounced its quantum coherence. Henry and Schred have had first-hand experience of this relation in their adventure.

Henry is Decohered by the Environment

CHAPTER 9

What is the Environment of Quantum Systems?

9.1 COHERENT OSCILLATIONS AND ENVIRONMENTAL DECOHERENCE

The discovery of a new crystal with special quantum properties has started a fierce struggle between the two rivals, Henry and Eve. They try to outrace each other by all means available to them, no hold bars, so as to get first to this new material that is deposited in an abandoned underground mine. Henry considers using his quantum rocket (Chapter 6) to get down there, grab the material and get up again quickly before Eve can intercept him, using the energy stored in the rocket as a propellant. However, as he has learnt from his Pisan adventure (Chapter 6), the amount of energy in the quantum rocket is highly unpredictable. The reason is that the rocket moves Henry to a superposition of states with different energies, followed by a measurement that causes *random* collapse to one of those states. Henry cannot know in advance which state that would be. Using such an unpredictable device in an underground mine is not a good idea.

On the other hand, as his mentor pointed out, the Split and Recombine functions in his suit perform a totally predictable coherent (unitary) process: the former splits Henry in two superposed states, while the latter recombines them. There is no uncertainty or measurement involved in these processes. Henry has thus come up with the idea of combining the energy transfer of the rocket and the coherent Split and Recombine functions. He has thereby created a new control button, named after the famed physicist I. Rabi, the discoverer of Rabi oscillations (Section 9.2).

In his descent into the mine, Henry resorts to the new Rabi button to perform a *coherent transfer* from the state where he is located outside the mine, to another state with different potential energy, in which he is located down the mine. At the outset, Henry is completely outside the mine; then he gradually, coherently, splits between being outside and inside the mine at the same time. Halfway through the process he is in an equal superposition of his outside and inside quantum versions. At the end of the process, Henry is entirely transferred to the state inside the mine.

Thus, the new Rabi button combines energy transfer, which enables Henry to change his *potential* energy level, and coherent splitting between the two states. The energy of the rocket allows him to go *down* or up—i.e., change his potential energy—as opposed to simply passing through two doors with the same energy, as he did when first testing his new quantum suit (Chapter 2).

If we analyze this process in more detail, we realize that not only a coherent superposition is in play here, but also *time-dependent interferences*. As opposed to the Split button, which splits Henry in two quantum versions in a single step, the new Rabi button performs a *continuous* process during which Henry's two versions (states) destructively interfere outside the mine and constructively interfere inside the mine, thereby gradually leading Henry's superposition state into the mine.

Henry counts on this process of Rabi *oscillation* to carry him first down and then up again. This is because he expects, by continuously pressing the control button, to oscillate back and forth between the outside and inside states. To simplify the control, Henry designs it in such a way that one push of the button would gradually transfer him from being completely outside to being (in due time) completely inside the mine. Another push of the button would slowly transfer him back outside, by a reversal of the same coherent process.

Yet something goes amiss in the process. The first push indeed successfully takes him down into the mine, where he snatches the quantum crystals. However, when he tries to go back up, he fails. Why so? The reason lies in one of the most fundamental processes in real quantum systems: *decoherence* due to interaction with the environment. In our story, the environment is represented by Eve's sensors around the mine to monitor what goes on. These devices unfortunately interact with Henry, as they "sense" him. Each device acts as if it *measures* Henry's position. However, since these are short-range, low-efficiency devices, none of them can individually measure Henry's exact location. Each device collects only a small amount of information regarding Henry's location by interacting with him. However, this interaction has an overall huge effect on Henry, for a reason discussed in Chapter 8: Henry's position gradually becomes entangled with the devices' positions. However, contrary to Henry's strong entanglement to Schred (in Chapter 8), Henry is now weakly entangled to many devices. Yet Schred, being an extremely smart cat, could undo the entanglement by reuniting with its other quantum versions. In Henry's current predicament there are so many of Eve's devices that they cannot all be disentangled from him, as this would require a huge number of disentangling operations! Thus Henry, slowly but inevitably, loses his identity by becoming part of an entangled state with the environment.

As Henry realized in his break-in to Eve's house (in Chapter 7), being part of an entangled state ruins the possibility of interfering with oneself. This realization now leads Henry to the conclusion that he cannot leave the mine by the same way that he came in: On his way out of the mine, the Rabi oscillation is supposed to split Henry in two quantum states—one still inside the mine and the other outside—which then *interfere* so as to transfer Henry outside. However, since Henry is now entangled to the sensors, he cannot undergo interference. Hence his entire coherent transfer back outside the mine cannot be completed, and he remains stuck halfway out of the mine. Fortunately for him, thanks to Johnny's friends' rough but decisive actions, Eve's sensors are disabled, thus disentangling Henry and allowing him to coherently exit the mine.

Let us recapitulate on Henry's quest for the quantum crystals. Henry has installed a new Rabi button which combines energy transfer from the quantum rocket to the Split button. This enables him to coherently transfer from one energy level to another—in this case, down the mine. This process requires quantum coherence, as it relies on quantum interference of Henry's states, both outside and inside the mine. Eve's sensors, by measuring their surroundings, are entangling to Henry slowly, as each device weakly couples to him. Since there are numerous sensors, Henry eventually becomes highly entangled to the multitude of sensors. This entanglement to the sensors, representing the environment, makes Henry lose his coherence and become classical, rendering his quantum superpowers useless. The effect of so many sensors cannot be simply undone, as disentangling each of them from Henry would be a tantalizing task. Instead, Henry's mentor and his friends actively disable the sensors to stop their interaction with Henry, thus effectively isolating Henry (the system) from the sensors (the environment). This isolation enables Henry to reactivate his Rabi button and leave the mine with his precious quantum crystals.

How does our story relate to real-world scenarios? As discussed in detail in Chapter 8, any interaction between the system—such as an atom or a photon—and the environment, consisting of many roaming atoms or errant photons, causes the system and the environment to become entangled. The interaction with each constituent of the environment is weak, causing very limited entanglement. Yet together, these many constituents and the quantum system become completely entangled, making the system lose its quantum coherence; i.e. causing *environment-induced decoherence*.

But is there a clear distinction between an *environment* and a *system*? Where does their dividing line pass? In Henry's adventure the distinction is clear: Henry is the system, whereas Eve's sensors are the environment. However, in real experimental setups the dividing line is defined by *controllability*. A system stands for the collection of degrees of freedom that can be controlled; e.g. split, recombined or otherwise manipulated. The environment, by contrast, includes all the degrees of freedom that the experimenter cannot control or manipulate. In Chapter 8, Schred, while being entangled with Henry and thus jeopardizing Henry's break-in, was still considered part of a complex Henry–Schred system, and not part of the environment, because it could be "manipulated" to make it recombine with itself, thus reverting Henry to a quantum-coherent superposition.

In real experiments the energy levels of an electron inside an atom play the role of Henry's potential energy states; i.e. the lower and the higher electron-energy levels are analogous to Henry being inside or outside the mine, respectively. Henry's Rabi oscillation button is implemented by a laser that transfers energy coherently to the electron. Then, the electron oscillates between its excited (high-energy) and ground (low-energy) states. If one turns off the laser in the middle of the process, one creates a superposition state of the electron being both excited and unexcited at the same time. In fact, Rabi oscillation between electronic energy levels is one of the common experimental methods of creating a quantum superposition state.

9.2 DECOHERENCE AND DECAY IN AN ENVIRONMENT (A "BATH")

In his previous adventure, Henry's coherent (Rabi) oscillation between his two energy states was hampered by the cumulative effect of many tiny localizing sensors scattered around by Eve. This multitude of sensors is meant to represent some effects of the environment that will henceforth be called a "bath" in its quantum-mechanical description, for reasons to be clarified. The many degrees of freedom (DOF) of such a bath are assumed to interact with the system of interest (here, Henry) and possibly become entangled with it, as did Henry and Schred in Chapter 7. As in Chapter 8, such entanglement, upon averaging over (tracing out) the bath DOF, *decoheres* a quantum superposition state of the system.

This consensual narrative on the origin of decoherence conceals certain puzzling and disconcerting issues, as we will show. To introduce the subject, let us first discuss the main features of baths and quantum system–bath interactions.

Typical solid-state baths are composed of a huge number of tiny oscillators that can be visualized as nanoscopic springs, whose quantum mechanical oscillation modes are called "phonon modes". These modes have almost continuously distributed energies $\hbar\omega$ and momenta $\hbar k$. The dependence of ω on k is called the phonon spectrum (Figure 9.1)—a characteristic of the solid-state material in question.

On the other hand, the number of *excited oscillators or quanta* (phonons) with given $\hbar\omega$, $\hbar k$ is determined in a unit volume by the temperature T of the material. This unique dependence of the number of quanta per unit volume on T, and not on the size of the material chunk or on some other fleeting changes, is a signature of a thermostat, better known as *thermal bath*. A bath is presumed to be so large that its temperature is fixed, regardless of any microscopic/nanoscopic fluctuations within it; e.g., its interactions with sparse quantum systems—for example, impurities such as quantum dots implanted in the solid (Figure 9.2).

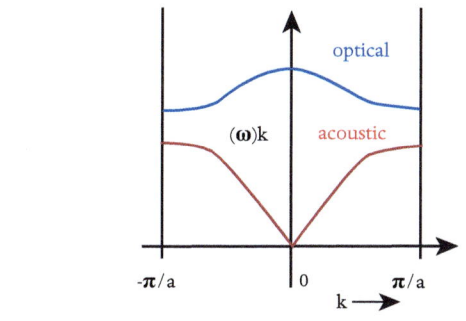

Fig. 9.1 Phonon spectrum in crystals with atom distance a. Upper (lower) branch - optical (acoustic) phonons.

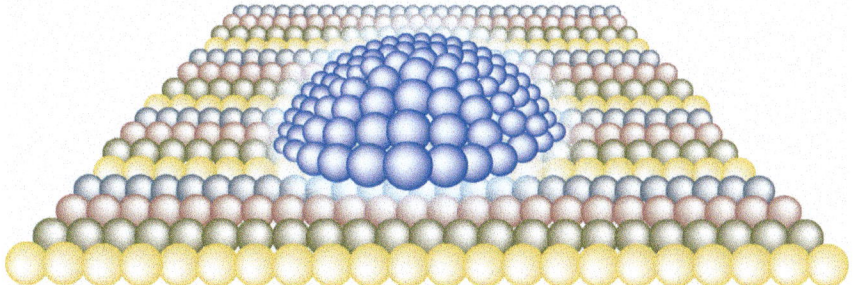

Fig. 9.2 A quantum dot composed of many impurity atoms that act as one big artificial atom (blue) in contact with the bulk of the solid composed of yellow and green atoms that act as a bath.

A similar bath is that of electromagnetic (EM) quanta or photons. This bath permeates all empty space, which is populated by microwave photons emanating from the ubiquitous cosmic background (Big Bang relic) radiation at 2.7 K, but also by photons at all possible frequencies from diverse cosmic sources. There are also artificial, finite, EM baths engineered in cavities with a designed spectrum which determines their interactions with atoms in the cavity (Figure 9.3).

Another commonplace bath is a "buffer gas" acting as a thermal bath, as its noble- or rare-gas atoms randomly collide with a "target gas" atom or molecule that constitutes a system of interest. The rate of collisions and the occupancy of energy states in the buffer gas depend on its pressure and temperature (Figure 9.4).

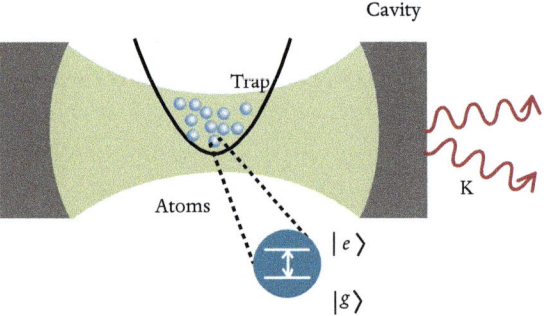

Fig. 9.3 Atoms trapped in a light-cavity. The atom, whose internal levels e and g are separated by interlevel energy $\hbar\omega$, interacts with a bath of photonic modes in the cavity that are near-resonant with the atomic interlevel energy.

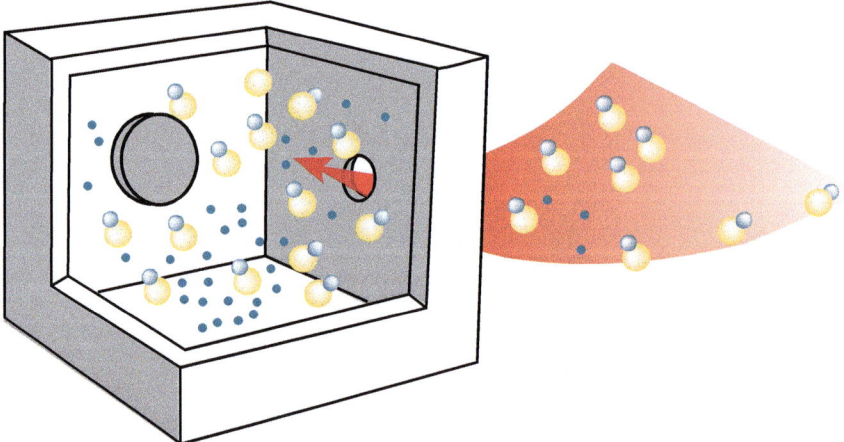

Fig. 9.4 Buffer gas bath in a cell. Random collisions of target gas molecules (yellow) with noble-gas atoms, serving as buffer-gas (blue).

The foregoing descriptions of baths and system–bath interactions have their origin in the statistical (classical) theory of gases by Maxwell, Boltzmann and Gibbs in the nineteenth century (Chapter 8). Throughout the 1930s to the 1950s they were transformed to the realm of QM by many outstanding physicists, notably L. Landau (USSR), F. Bloch (USA), V. Weisskopf, E. Wigner, H. Bethe (Germany, then USA), R. Kubo (Japan), U. Fano, M. Zwanzig and P. Anderson (USA).

The pertinent question here is: under what conditions and by what means do such interactions entangle the system and the bath, thereby decohering the system state? The answer is that one of the following effects must come to pass.

1) Quanta exchange near thermal equilibrium

The system exchanges quanta of energy $\hbar\omega$ with the bath, intermittently absorbing and emitting elementary bath-excitations (e.g. phonons or photons). The system and the bath eventually reach a thermal-equilibrium state in which the respective rates of these two processes are "balanced" at the bath temperature T (Figure 9.5). These are random, incoherent processes. Quantum-mechanically this randomness comes about because (Sections 8.1, 8.2 and 9.1) the bath states involved in these processes are averaged over (traced out). Classically, such randomness is obtained if we assume random phases of the bath oscillators. The result is the same: the destruction of coherent oscillations in the system by the bath—a common type of decoherence that was termed *proper dephasing*.

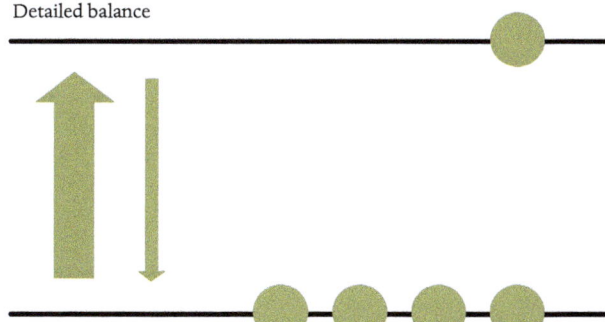

Fig. 9.5 Instantaneous ("detailed") balance between absorption and emission rates of quanta in a two-level system (TLS) at temperature T. Arrow thickness indicates the rate. The number of circles represents the population or occupation probability of each level. Here, the lower-level population is four times larger than that of the upper level.

This decoherence is deemed to be *irreversible* because, in order to undo it, we presumably need to know the phases of both the system and the bath oscillations, which contradicts their assumed randomness. As shown in Chapter 12, there is a way to circumvent this irreversibility. However, let us accept for the time being its compelling logic, because it agrees with the requirements of thermodynamics, as we explain. The system is assumed at a given time to be in a coherently oscillating state—namely, out of thermal equilibrium (since at thermal equilibrium the state is thermal, i.e. completely phase-random). According to the universal second law of thermodynamics, put forward by R. Clausius (Germany) around 1850, an out-of-equilibrium system must evolve towards equilibrium by increasing the total entropy of the system and the bath combined, but since the bath is too large to change, being always in equilibrium, only the system entropy grows. In thermodynamic terms this process is in fact *heat exchange*, defined as concurrent *exchange of energy and entropy* between the system and the bath. Thus the elementary acts of emission and absorption discussed previously must give rise to irreversible decoherence.

2) Spontaneous emission

The picture changes when the system is in contact with an empty bath; i.e., at $T = 0$. In the absence of quanta to be absorbed, the system can only emit quanta into the bath, provided the excited state of the system is initially populated with some probability (Figure 9.6). This "spontaneous emission" is a uniquely quantum-mechanical effect. A classical excited system, such as an oscillator with an appreciable energy, albeit unstable, would not give away its energy to an empty bath unless pushed by an external force. By contrast, quantum-mechanically, spontaneous emission causes the populations of excited states of the system to decay to the lowest (ground) state. Whenever such decay changes and distorts the ratio of probability amplitudes in a superposition state of the system, the outcome is decoherence, as in the case of finite temperature discussed previously. Yet in general, decay and decoherence do not have to go hand-in-hand.

3) Bath-induced energy shifts

The system energy levels may be appreciably shifted by the interaction with the bath, but without changing the level populations. Such shifts may be the

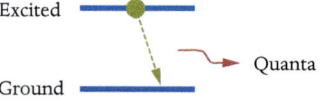

Fig. 9.6 Spontaneous emission of quanta by an initially excited two-level system.

outcome of soft collisions of the system with the bath quanta, whose energies are far-from-resonance with transitions between the system levels, which suppresses the energy exchange between the bath and systems and thereby the population decay, but may perturb the energy levels of the system. Since different energy states of the system may be shifted by different amounts by the bath, a superposition of such states may accumulate random phase shifts that are the products of the respective energy shifts and time, thereby leading to decoherence (Figure 9.7).

Bath-induced decoherence is truly ubiquitous. It is nearly impossible to find quantum systems so isolated from their environment that they practically "live forever" in coherent-superposition or multipartite-entangled states. The relevant lifetime of such states is normally too short to allow for any desirable task that relies on their coherence or entanglement, unless special measures are taken to extend this time (Chapter 10).

By contrast, bath-induced decay of excited states of a quantum system is often a lesser obstacle towards the implementation of desired tasks. Thus the decay of unstable levels of radioactive materials, which is a form of spontaneous emission that proceeds through coupling to several channels or baths, is unavoidable (Chapter 11), but may end up in metastable levels that live long enough for our purposes. The same is true for spontaneous emission of optical photons by electronically excited atoms and molecules, and such emission may shelve them in long-lived states. An even longer lifetime characterizes atomic or molecular levels that spontaneously emit EM radiation with much lower frequency, such as infrared or microwave photons, or vibrationally excited states in solids that decay via phonon emission. Nevertheless, there is strong impetus for manipulating the lifetime of any decaying state of a quantum system. While such an undertaking would have been impossible until recently, there are now good prospects for its feasibility, as shown in Chapter 10.

The foregoing picture of bath-induced decoherence and decay has been confirmed by a multitude of experiments and calculations. Yet an important

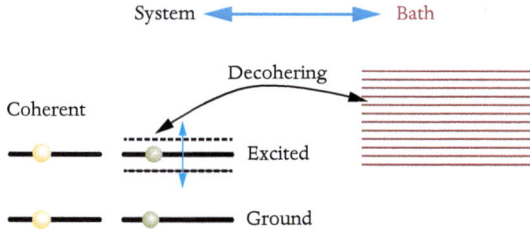

Fig. 9.7 Bath-induced energy shifts randomize the system level distance and thus cause decoherence.

conceptual hurdle remains: where does one draw the line between the system and the bath? We shall discuss this issue in Section 9.3, but an example may clarify its disconcerting nature. Consider the composite system consisting of an atom interacting with cavity photons (Figure 9.3). The atom is commonly viewed as an open, irreversibly decaying and/or decohering system, and the cavity photons as an EM bath that may very weakly couple to the outside world—another infinite EM bath. Alternatively, the composite system may be treated as nearly closed (apart from weak coupling to the outside)—a "dressed atom" comprising all the atom and photon degrees-of-freedom (DOF). As with any closed, unperturbed system its "dressed" state must keep the same purity or entropy at all times, and its overall energy must not change. Yet the difficulty of keeping track of all DOF may be prohibitive, as the boundary between irreversibility and reversibility is not rigid, but a matter of practicality.

9.3 IRREVERSIBILITY AND THE ARROW OF TIME

The unitarity of QM ensures its symmetry under time reversal: by running the evolution backwards in time, the final and initial states are interchanged, and so are our past and future. On the other hand, decay and decoherence are manifestations of *irreversible* evolution of "open" quantum systems that are in contact with a bath. Thus, a coherently-superposed state of an open system that has been corrupted by bath-induced decoherence or decay cannot be brought back by time-reversal, because the process is *non-unitary*, involving the tracing out of or ignoring the bath (Sections 8.4 and 9.2). This difference between the evolution of "closed" and "open" systems is mathematically clear but physically clouded; if all that matters for this difference is which bath states are to be ignored, then the division between system and bath DOF becomes a matter of expediency. As our experimental and calculational capabilities improve (Chapter 15), we shall be able to include more and more bath modes and decay/decoherence channels in our quantum system of interest, thereby turning it from "open" to "closed".

W. Zurek (USA) has, over the past few decades, been making an even stronger case for the role of the bath in inducing irreversibility. According to Zurek, the environment acts as a *meter* rendering coherence inexistent on large space- and time-scales: the more complex the object, the faster its quantum state decoheres. Yet Zurek's view faces objections similar to those raised above. Zurek assumes that the environment/bath chooses the mixture of long-lived ("pointer") states into which the initial state decoheres, but there is often a great deal of arbitrariness in such a choice, which is far from being universal.

Whatever the resolution of this conundrum, a more fundamental objection may be raised against the consensus that *irreversibility, resulting in thermal equilibration, is inherent in closed many-particle quantum systems, despite their unitarity.* This consensus can be traced back to Von Neumann's enigmatic ansatz (i.e., assertion without proof) in 1929, which has since been termed the *Eigenstate Thermalization Hypothesis* (ETH). According to the ETH, a typical observable must eventually become indistinguishable from its thermal-equilibrium counterpart, regardless of the initial state. How can seemingly reversible evolution yield irreversibility? A possible way to vindicate the ETH, as shown by M. Srednicki (USA) in the 1990s, is to assume that our observational limitations amount to *coarse-graining* of the unitary evolution; that is, we cannot probe the tremendously fast oscillations between coherently superposed states of highly complex systems. In macroscopic systems, such oscillations may have periods *near the ultimate limit for the existence of time*—the *Planck time* (Chapter 6)—so that they may be meaningless, whatever our experimental capabilities. Surprisingly, even a moderate number of particles allows us to ignore such coherent oscillations, as shown by recent experiments and calculations. It thus appears that the ETH—this incredibly bold ansatz by Von Neumann—is universally valid.

Yet there is a notable exception to the ETH whereby quantumness is revealed at long times even in highly complex systems. If the energies $\hbar\omega_n$ of the states of such a system that are enumerated by integers $n = 1,2,\ldots N$ are all multiples of each other (are *commensurate*, in mathematical jargon), then at certain times the time-dependent phases associated with the oscillations of such states in any initial superposition they form become the same (Section 9.4, Appendix); i.e., the initial superposition state is *revived* and reversibility is restored, at least momentarily. Such behavior shows that quantumness or unitarity lurks beneath the façade of seemingly equilibrated quantum systems, contrary to their widely assumed irreversibility. Recent experiments by J. Schmiedmayer (Austria) on thousands of ultracold atoms trapped in a box-shaped potential, which allows for commensurate quantized energies, have confirmed this spectacular quantum effect.

Although such revivals are of quantum origin, they are akin to the general type of revivals predicted by H. Poincaré (France) in 1895 for arbitrarily large systems of classical particles. Since classical (Newtonian) mechanics possesses symmetry under time reversal (just like QM), Poincaré argued that inevitably a time will come when the positions and momenta of all particles are the same as they were initially; i.e., the overall state will be revived. This argument was used by Poincaré to discredit Boltzmann's H-theorem, whereby entropy (disorder) in a large system of gas particles will grow until thermal equilibrium is reached, as a microscopic corroboration of the second law of thermodynamics. Not only

Poincaré, but also J. J. Loschmidt, Boltzmann's friend and colleague, rejected the H-theorem by advancing the idea of an "echo": whatever arguments support entropy growth, they can be reversed if the particles are all made to turn around on their trajectories. Boltzmann retorted: "Why don't you try to do that?", implying that such a feat would be impossible. Yet in fact, as shown in Chapter 12, Loschmidt's echo proved to be feasible, at least in some systems.

As a result of Poincaré's and Loschmidt's objections, the consensus at the turn of the nieteenth century turned against Boltzmann's theory of gases, which may have triggered his suicidal mood in 1905. His tragic death might have been avoided had he known that in that same year Einstein published the theory of Brownian motion, which took Boltzmann's gas theory for granted. Still, it is now clear that the growth of entropy with time of a large system, be it classical or quantum, prior to its equilibration, cannot be deduced without an assumption of coarse-graining over experimentally inaccessible time-scales.

The foregoing arguments "pro and con" irreversibility are just glimpses into the subtlety of this fundamental issue. In the 1930s the tide was turning in favor of irreversibility when A. Eddington (UK) coined the expression "the arrow of time" to indicate that thermodynamics and statistical physics support this assertion, as does our daily experience: spilled milk does not gather itself into the mug, broken glass does not recreate a bottle, and so on. Yet can such unlikely processes be ruled out *in principle*, and, if so, why? This foundational question is still being debated.

The issue at hand is not merely academic: it may be literally a matter of life and death. The transition from life to death in all organisms is describable by entropy increase as a result of the cessation of metabolic activities. Are there quantum features in such biological processes, just like in the decay processes reviewed above?

At present we cannot monitor metabolism *in vivo* at the quantum level, so that no conclusive answer can be given now. Still, the (somewhat speculative) surmise that quantumness may be involved in biology is intriguing. If substantiated, it would imply that organisms, in the course of evolution, may have developed forms of quantumness that are resilient and not vulnerable to decoherence, in spite of the fact that a wet, warm environment—the epitome of a bath—is essential for metabolism.

Crumbling World

The fabric of the universe is wearing out
By the relentless course of decoherence,
Which spreads disorder all about,

Erasing every shred of interference.
Ah, time arrow is ticking fast.
The world is aging, with no hope in sight,
Since quantumness is doomed: decay it must!
But could the clock be stopped? It might!

9.4 APPENDIX: COHERENT (RABI) OSCILLATIONS AND DECAY

This appendix, building upon previously encountered concepts, mathematically describes coherent (Rabi) oscillations in a driven quantum system, and the decay of such oscillations, caused by entanglement of the system to the environment, as illustrated in Henry's present adventure.

The new Rabi button that Henry has added to his amazing quantum suit extends the previous Split and Recombine buttons, in that it also integrates the quantum rocket operation. As we recall from Chapter 6, the rocket action allows the transfer of energy, which is mandatory for Henry's present purpose, but also constitutes a *dynamic process* that *gradually* changes Henry's state. For this purpose, as in Section 6.4, we must use the famous Schrödinger equation, which describes the dynamical change of quantum states:

$$H \,|\, \psi >= i\hbar \,(d|\psi >) \,/dt$$

In the present case, the Hamiltonian, which describes the operator that changes the dynamics, is the Rabi operator, $H = V\sigma_x$, where V is a number representing the magnitude or strength of the operation, and σ_x is the x-Pauli operator that has the (by now familiar) matrix form, $\sigma_x = \begin{pmatrix} 0 & 1 \\ 1 & 0 \end{pmatrix}$. As it acts on the basis of $|\uparrow>, |\downarrow>$, the states of Henry above the ground (henceforth surface state) and in the mine (henceforth mine state), respectively, the Rabi button moves Henry from his surface state to his mine state. Henry's state in the Schrödinger equation in general extends over the two possible locations:

$$|\,\psi\,(\mathbf{t})\,>= a_1 \uparrow (t)\,|\uparrow> + \,a_1 \downarrow (t)\,|\downarrow>$$

The time-dependence of Henry's state is needed, since we are discussing the dynamics of Henry's attempt to go down the mine. Plugging the Hamiltonian and Henry's most general state in the Schrödinger equation results in:

$$Va_\uparrow(t)\,|\downarrow> + \,Va_\downarrow(t)\,|\uparrow>= i\hbar\frac{da_\uparrow(t)}{dt}\,\bigg|\uparrow> + \,i\hbar\frac{da_\downarrow(t)}{dt}\,\bigg|\downarrow>.$$

Here, the Hamiltonian operation, ruled by σ_x, has exchanged the states $|\uparrow> \Longleftrightarrow |\downarrow>$ on the left-hand side.

The dynamical equation above for the probability amplitudes of all the states (here, two states) appears complicated. The crucial new insight in this appendix is that this cumbersome, composite equation can decomposed into simple dynamical equations—one for each *orthogonal state*. Thus the equation above results in the following set of coupled equations:

$$Va_{1\uparrow}(t) = i\hbar\,(da_{1\downarrow}(t))\,/dt \qquad \text{corresponding to } |\downarrow> \text{ (mine state)}$$
$$Va_{1\downarrow}(t) = i\hbar\,(da_{1\uparrow}(t))\,/dt \qquad \text{corresponding to } |\uparrow> \text{ (surface state)}$$

As can be seen, the two equations are coupled and therefore need to be solved together. By differentiating the upper equation and using the lower one, we obtain:

$$\frac{d}{dt}Va_{\uparrow}(t) = \frac{d}{dt}i\hbar\frac{da_{\downarrow}(t)}{dt}$$
$$V\frac{da_{\uparrow}(t)}{dt} = i\hbar\frac{d^2a_{\downarrow}(t)}{dt^2}$$
$$V\frac{V}{i\hbar}a_{\downarrow}(t) = i\hbar\frac{d^2a_{\downarrow}(t)}{dt^2}$$

This finally results in the second-order differential equation (keeping in mind that $i^2 = -1$):

$$\left(\frac{V}{i\hbar}\right)^2 a_{\downarrow}(t) = \frac{d^2a_{\downarrow}(t)}{dt^2}$$

$$-\left(\frac{V}{\hbar}\right)^2 a_{\downarrow}(t) = \frac{d^2a_{\downarrow}(t)}{dt^2}$$

The solution to this equation is an oscillatory function, of the sine or cosine form, depending on the initial condition: Since Henry starts on the surface, his state is given by the following Rabi oscillation:

$$|\,\psi\,(\mathbf{t})> = a_{1\uparrow}(t)|\uparrow> + a_{1\downarrow}(t)|\downarrow> = \cos(\Omega t)|\uparrow> + \sin(\Omega t)|\downarrow>$$

where the oscillation (Rabi) frequency of the probability amplitudes is given by $\Omega = V/\hbar$. Henry's state at $t = 0$ is entirely a surface state when $\cos 0 = 1$. It gradually turns into a fully superposed state of being equally in the surface and mine states at time $\Omega t = \pi/4$, since then $\cos \pi/4 = \sin \pi/4 = \sqrt{1/2}$. The state corresponds to being fully in the mine at $\Omega t = \pi/2$: then $\sin \pi/2 = 1$, $\cos \pi/2 = 0$. If Henry were to continue his Rabi oscillation over an equal lapse of time, he would return to the surface (with the opposite sign before the state). This Rabi oscillation represents *coherent* dynamics, which is both deterministic and *oscillatory* in the sense that it periodically toggles between the two states.

The essential physics behind this *coherent* oscillation is described by this diagram.

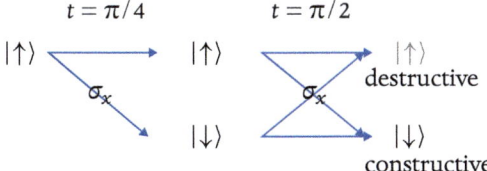

According to this diagram, since Henry starts on the surface he gradually becomes a superposition of the surface and mine states under the action of the Rabi operator. Then, the same Rabi operator, acting on this superposed state, flips the two states so that surface→mine *and* mine→surface. The question arises: why does Henry not go back up as a result of this flip? The answer is: because of the *interference* of his two-state amplitudes that have either the same or the opposite signs. There are two channels involved in this dynamic process: part of the *surface Henry* stays there while the other part goes down the mine; likewise, part of the *mine Henry* stays below while the other part goes to the surface. These two channels interfere, since *they involve the same states*. During the downward phase of the oscillation there is destructive interference in the *surface Henry* channel and constructive interference in the *mine Henry* channel, and this destructive interference moves most of Henry's amplitude to the mine. If Henry could continue his oscillation, a reversal of the processes above would happen: destructive interference in the channel would become constructive interference, and vice versa, in the other channel.

Now let us examine Henry's attempt to go back up, after he has found the quantum crystal. Without Eve's tampering with his action, Henry could just reactivate the Rabi button, resulting in his coherent transfer back up. However, Eve's sensors have already started to entangle with Henry, thereby affecting his dynamics, as will be explained. These sensors have the states $|0>$ that signifies no detection, and $|1>$ that signifies detection.

Let us first consider that there is only one sensor. Then the complete state of Henry and the sensor is, in its most general form, given by:

$$|\psi(t)> = a_{1(\uparrow 0)}(t)|\uparrow>|0> + a_{1(\downarrow 0)}(t)|\downarrow>|0>$$
$$+ a_{1(\uparrow 1)}(t)|\uparrow>|1> + a_{1(\downarrow 1)}(t)|\downarrow>|1>$$

We know write the full dynamical (Schrödinger) equation for this coupled state, considering that the Rabi operator acts only on Henry's state:

$$Va_{1(\uparrow 0)}(t)|\downarrow>|0> + Va_{1(\downarrow 0)}(t)|\uparrow>|0> + Va_{1(\uparrow 1)}(t)|\downarrow>|1>$$
$$+ Va_{1(\downarrow 1)}(t)|\uparrow>|1> = i\hbar\left(da_{1(\uparrow 0)}(t)\right)/dt|\uparrow>|0>$$
$$+ i\hbar\left(da_{1(\downarrow 0)}(t)\right)/dt|\downarrow>|0> + i\hbar\left(da_{1(\uparrow 1)}(t)\right)/dt|\uparrow>|1>$$
$$+ i\hbar\left(da_{1(\downarrow 1)}(t)\right)/dt|\downarrow>|1>$$

There are now *four* sets of equations—one for each *orthogonal state*:

$$Va_{1\,(\uparrow 0)}(t) = i\hbar \left(da_{1\,(\downarrow 0)}(t)\right)/dt \qquad \text{corresponding to } |\downarrow\rangle|0\rangle$$
$$Va_{1\,(\uparrow 1)}(t) = i\hbar \left(da_{1\,(\downarrow 1)}(t)\right)/dt \qquad \text{corresponding to } |\downarrow\rangle|1\rangle$$
$$Va_{1\,(\downarrow 0)}(t) = i\hbar \left(da_{1\,(\uparrow 0)}(t)\right)/dt \qquad \text{corresponding to } |\uparrow\rangle|0\rangle$$
$$Va_{1\,(\downarrow 1)}(t) = i\hbar \left(da_{1\,(\uparrow 1)}(t)\right)/dt \qquad \text{corresponding to } |\uparrow\rangle|1\rangle$$

Upon examining these equations, it becomes clear that there is *no coupling between* the $|0\rangle$ and $|1\rangle$ states. There are several cases to reckon with as a result of this lack of coupling. The first is that if Henry is completely undetected—i.e., starts being correlated with $|0\rangle$ and continues this way—the case becomes identical to his downward transition, and he can then coherently go back up.

Another case of interest occurs if Henry and the sensor suddenly become entangled, while Henry is in a fully superposed state. Namely, just after $\Omega t = \pi/4$, the detector flips from 0 to 1, only for the *surface Henry*. This "detection" caused by entanglement to the sensor is described by the following change of the complete state:

$$\sqrt{(1/2}\,|\uparrow\rangle|0\rangle + \sqrt{(1/2)}\,\downarrow\rangle|0\rangle = \sqrt{(1/2)}\,|\uparrow\rangle|1\rangle$$
$$+ \sqrt{(1/2)}\,|\downarrow\rangle|0\rangle$$

If Henry then continues to press the Rabi button, a dramatic occurrence unfolds: Henry's *surface and mine* versions become *completely decoupled*. In other words, they evolve separately, since, due to the entanglement of Henry and the sensor, his two versions *do not interfere*. The state at $\Omega t = \pi/2$ becomes:

$$|\psi\,(t=\pi/2\Omega)\rangle = 1/2\,|\uparrow\rangle|0\rangle + 1/2\,|\downarrow\rangle|0\rangle + 1/2\,|\uparrow\rangle|1\rangle$$
$$+ 1/2\,|\downarrow\rangle|1\rangle$$

Here, due to the separate evolution of the detected and undetected parts, the first two terms on the right-hand side arise from the evolution of the undetected $|0\rangle$ part, and the last two terms appear due to the evolution of the detected $|1\rangle$ part. Henry now has a 50% chance of being both on the surface and in the mine, instead of 100% certainty of being on the surface if he were disentangled from Eve's sensors. The entire process is shown in the following diagram:

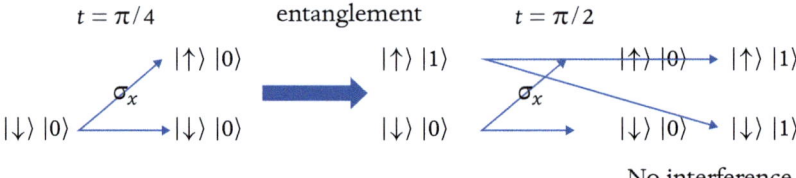

$t = \pi/4$ entanglement $t = \pi/2$

No interference

To summarise this subsection we have shown that *coherent* transfer between two states of a quantum system, caused by their external driving Hamiltonian, results in a Rabi oscillation, a sine or a cosine as a function of the transfer time. This oscillation exhibits interference between two channels corresponding to two different outcomes: staying in the initial state or transferring to the other, orthogonal state. This process completely changes if the system (here, Henry) becomes entangled to an *outside system*, which is henceforth called the *environment* (here, the sensor). Then, coherence disappears and interferences between channels cannot ensue. Such a process, caused by interaction with the environment, is called *decoherence*.

In Henry's case, instead of one sensor that strongly interacts with him, which would lead to their complete entanglement, there are many weakly interacting sensors, each one of them becoming only slightly entangled with Henry. What complicates this scenario is that after the weak entanglement there are many *uncoupled* channels, one for each sensor, each evolving separately.

To explore the consequences of such a multichannel scenario, we introduce a very important game changer, which has been mostly ignored thus far (except in Chapter 6); namely, the *energy difference* between the states involved. Let us assume that Henry's states of being above and below the ground have different energies, and take the energy of above-the-ground Henry to be $E_{above} = \hbar\omega_a$ in excess of the energy of below-the-ground Henry which is chosen to be 0. Then, the Hamiltonian, which represents the energy of the entire system, is given by:

$$H = \hbar\omega_a \, |\uparrow><\uparrow|$$

Again applying Schrödinger's equation results in:

$$\hbar\omega_a a_\uparrow(t) \left|\uparrow> = i\hbar\frac{da_\uparrow(t)}{dt}\right| \uparrow>$$

The solution to this simple differential equation is $a_\uparrow(t) = e^{-i\omega_a t}a_\uparrow(0)$. This result shows that the excess energy of the upper (surface) state simply introduces an *oscillating phase* into the state's probability amplitude. Since $|e^{-i\omega t}| = 1$, this means that the *probability* is not affected by the oscillation.

Let us revisit the scenario wherein Henry is starting to be weakly coupled to a plethora of sensors, now assuming that each of them has its own energy level. These sensors are represented by the states $|0>_j$ and $|1>_j$ for sensor j with energy $\hbar\omega_j$. Since all the sensors are so weakly coupled to Henry, the chance that two or more sensors detect him is extremely small and will be neglected henceforth. The complete state of Henry and the sensors is then described by:

$$| \psi(t) > = a_{\uparrow 0}(t) e^{i \omega_a t} |\uparrow>|\, 0 > + a_{\downarrow 0}(t) |\downarrow>|\, 0 >$$
$$+ \sum_j a_{\uparrow j}(t) e^{i(\omega_a + \omega_j)t} |\uparrow>|\, 1>_j + a_{\downarrow j}(t) e^{i \omega_j t} |\downarrow>|\, 1>_j.$$

The interaction between Henry and the sensors can be represented by a simple Hamiltonian, akin to the σ_x Hamiltonian from the previous section, which causes the transfer from one state to another, while adhering to *energy conservation*: i.e., if a sensor is excited to its higher-energy state, then Henry must lower his energy to $|\downarrow>$, and vice-versa. It has the form:

$$H_{int} = \sum_j \mu_j \left(|\downarrow><\uparrow|\, |0>_j < 1| + |\uparrow><\downarrow|\, |1>_j < 0| \right).$$

Here, μ_j is the (small) interaction strength between the jth sensor and Henry. The first term on the right-hand side indicates that the surface Henry loses his energy to is his mine counterpart, while the sensors detect him and thereby climb to their higher energy state. The second term on the right-hand side has the opposite effect—an increase in Henry's energy to the surface state—while at the same time a sensor loses its energy. This second term is mandatory for the Hamiltonian to be unitary; i.e., have no preference for one direction of the transfer over its opposite.

By this interaction, energy is transferred from one system to another—here from Henry to the sensors, or vice versa. As indicated by the summation over j, there are many channels for this energy transfer—as many as the sensors.

From the Schrödinger equation we then arrive at the following set of equations:

$$\frac{da_{\uparrow 0}(t)}{dt} = -i\hbar^{-1} \sum_j \mu_j e^{i(\omega_a - \omega_j)t} a_{\downarrow j}$$

$$\frac{da_{\downarrow j}(t)}{dt} = -i\hbar^{-1} \mu_j e^{-i(\omega_a - \omega_j)t} a_{\uparrow 0}.$$

Notice that $a_{\downarrow 0}$ and $a_{\uparrow j}$ are excluded, because of energy conservation: $a_{\downarrow 0}$ has zero energy, which cannot change, while $a_{\uparrow j}$ has the maximal energy $\hbar(\omega_a + \omega_j)$ and no channel to lose it to. In contrast to the coupled equations in the previous scenario, here there are many time-dependencies in the equation. Notwithstanding this multitude, we can *integrate* the second equation and plug it into the first one. This results in:

$$\frac{da_{\uparrow 0}(t)}{dt} = -\int_0^t dt'\, e^{i\omega_a(t-t')} \sum_j \hbar^{-2} \mu_j^2 e^{-i\omega_j(t-t')} a_{\uparrow 0}(t').$$

The result is a generally insoluble integro-differential equation. We will therefore make several simplifying assumptions in order to solve it. The first assumption is that $a_{\uparrow 0}(t')$ changes only *slowly* over time, which stems from the fact that the sensors are *weakly* coupled to Henry. This assumption allows us to move $a_{\uparrow 0}(t')$ out of the integral and thereby obtain this extremely simple equation:

$$\frac{da_{\uparrow 0}(t)}{dt} = -R(t)a_{\uparrow 0}(t)$$

where

$$R(t) = \int_0^t dt' e^{i\omega_a(t-t')} \sum_j \hbar^{-2} \mu_j^2 e^{-i\omega_j(t-t')}$$

The first equation is now a simple *decay* equation; i.e., the probability amplitude $a_{\uparrow 0}(t)$ simply decays (reduces to zero) at the rate $R(t)$. While $R(t)$ still appears to be complicated, the message of this analysis is rather simple. Whenever Henry attempts to go back up to his surface state, whose probability is given by $|a_{\uparrow 0}(t)|^2$, his *energy-conserving* interaction with the many sensors sends him back down the mine at the rate $R(t)$. In other words, the weakly coupled sensors make Henry *decay* back to his lower-energy state. Hence, the many ways for Henry to lose energy will inevitably result in his "decay" back to the mine. We will discuss the structure of $R(t)$ and its meaning in later chapters.

In the real world, the environment is made of numerous microscopic or nanoscopic systems which incessantly become entangled with the quantum system at hand. This weak entanglement destroys coherent processes that may otherwise occur within the system, thereby decohering it and rendering it no longer "quantum". In effect, maintaining a quantum system in a coherent, excited state (with energy above the ground state) is extremely difficult, because the interaction of the system with the uncontrollable environment causes the system to decay to its lower-energy state, as we have shown.

Can Quantum Measurements Prevent Change?

10.1 THE WEDDING THAT NEVER HAPPENED

Henry's ordeal in the mine has shaken him profoundly: To his dismay, his glorious quantum suit has been rendered useless by Eve's sensors! Henry expresses to his mentor, Johnny, his deep fear that his quantum suit is done for, now that Eve has the better of him. His position, the observable protected by his quantum suit, loses its quantumness when exposed to her gadgets—either Eve's sensors in the mine or her surveillance cameras through which she tried to track him down in the subway as he was delivering Bob's briefcase (Chapter 4). "True, but only in part", his Mentor retorts calmly. "There are significant differences between the two incidents. The sensors in the mine were *weakly entangled* with you, and therefore *gradually disrupted the dynamic process* of your coherent transfer out of the mine, which as you know is a Rabi oscillation in time. By contrast, the cameras in the subway *abruptly collapsed* your coherent-superposition state at random to one or another of the superposed states. Try to get to the bottom of this difference; therein lies the solution to your problem. I can only offer a tip: have you heard of Zeno's arrow paradox and its quantum version?" "No, I haven't", Henry sighs exhaustedly. "Allow me to look it up and sleep on your riddle, Johnny."

After a sleepless night of extensive reading and pondering on Zeno's paradoxes and their quantum counterparts (which we recount later), Henry collapses into a bewildered slumber. A strange dream stirs his sleep, vexing his thoughts and evoking his troubled past: He is about to wed Eve, with whom he is apparently head-over-heels in love. "How did that happen?" he tries to recall. As expected in this state of bliss, he is coherently transported towards Eve down the aisle. But just as a small portion of him arrives next to her, that portion disappears in a flash of light! What remains instead is himself as far as he initially was from his bride.

In Henry's dream he approaches the bride via a *coherent process*, similar to his Rabi oscillation in Chapter 9. To recall, this oscillation is a coherent transfer between two states that are in this case separated in space. This continuous, dynamical process has a characteristic duration (time period). During the initial stages of this process, only a small portion of the quantum Henry is in the happily married state, next to Eve. Then, suddenly, the married quantum version of Henry disappears.

Henry is distraught and heart-broken. Instead of coherently propagating towards his bride, he has suddenly collapsed to a classical, single (pun intended) Henry. What is even stranger is that the light flash that has preceded his collapse has come from Eve's snapshot, but it must have come from another manifestation of Eve. Henry calls her evil Eve (EE), so much unlike his bride, sweet Eve (SE). "How on earth can there be two Eves simultaneously in this travesty of Dr Jekyll and Mr Hyde?" he wonders. "They act classically, so they do not form a superposition. Are they clones of each other? But QM forbids cloning, so the two Eves must be figments of my nightmare. But which one is closer to the real Eve?"

Henry knows that a snapshot is like a measurement that collapses a quantum superposition to one of the superposed states at random, with statistical weights determined by the respective probability amplitudes squared. For example, when EE measures a superposition of 1%-married Henry and 99%-single Henry, there is only a 1% probability of Henry collapsing into SE's loving arms, and a 99% chance that he will collapse to being single, which is indeed what he has experienced.

Henry keeps trying to be united in a state of matrimony with SE, while EE continues taking his pictures. As a result, a vicious circle or sequence ensues: Henry coherently shows up, but with low probability, near SE; EE takes his picture; Henry collapses to being far away from his bride; and this goes on and on. After what seem to be hours on end of EE's repeatedly taking his pictures, Henry sadly realizes in his dream: "SE and I can't get married as long as EE keeps taking her pictures!". He wakes up with this chilling thought of how his quantum present has intertwined with his classical past, in the form of the wedding ring that lies on his night table.

Oddly enough, Henry's nightmare indeed reflects the physical reality! The gist of it is Henry's and SE's continuous "rapprochement"—a coherent transfer of probability amplitude from one (single) state to another (matrimonial) state which is repeatedly interrupted by EE's snapshots. As in any coherent transfer, the probability amplitude of the final state builds up in time as a sine wave at the expense of the initial-state probability-amplitude, thus obeying the Rabi oscillation (Chapter 9). Since the probability of the final state is the square of its probability amplitude, it evolves as *a sine wave squared* if it is not interrupted. The transfer is then completed in half of the period of the sine wave squared, which we dub *transfer time* (Figure 10.1).

Henry's and SE's fate depends on how often EE takes his pictures compared to the *transfer time*. If the time-interval between pictures, which we dub *picture time, were longer than the transfer time*, so that the first picture would be taken after the transfer completion, Henry and SE would be happily married, because the snapshot would collapse Henry into his marital state with 100% probability.

Instead, EE chose the *picture time to be much shorter than the transfer time*, so that the first picture was taken early on in the coherent transfer process, when the transfer had only a very small probability. She then continued to take pictures at very short picture time intervals. EE did so deliberately, in order to take advantage of the fact that the sine-squared function that governs coherent transfer changes as the *time squared*, as long as this time is much smaller than the transfer time (Figure 10.1). Then, extremely short picture time renders the transfer probability to the final state proportional to the *squared ratio* between the picture time and the natural (uninterrupted) transfer time. This means that if the picture interval is decreased ten-fold, while the transfer time is unchanged, then the transfer probability is reduced a hundred-fold!

The squared ratio discussed above was the key to EE's success in blocking the transfer process, as we can now explain. When EE reduces the picture interval,

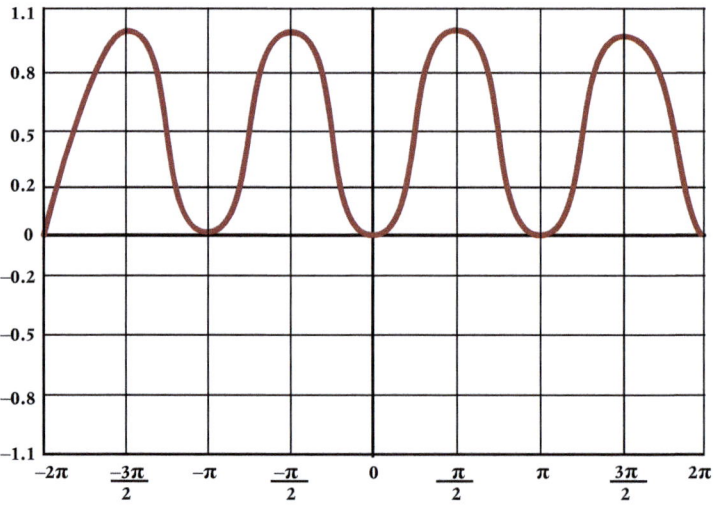

Fig. 10.1 Sine-squared function: amplitude dependence on phase.

the number of pictures during the complete transfer time increases, but this increase does not suffice to compensate for the decrease in the probability per picture that follows from the aforementioned squared ratio. Strikingly, the *overall* transfer probability (the product of the probability per picture and the number of pictures taken) then *decreases* as *the ratio between the picture time and transfer time*. Thus, the probability of Henry marrying SE becomes progressively smaller with the rate of EE's pictures!

This example illustrates the essence of what is known as the quantum Zeno effect (QZE). If a measurement collapses the quantum state, with a very high probability, to the initial state, then highly frequent repeated measurements can almost halt the coherent transfer process; i.e., nearly stop the change of the quantum state. Even though there is always a chance of transfer, its probability decreases drastically as the measurement rate increases.

In Section 10.3 we explain the origin of the name of this effect: it was meant to evoke the arrow paradox put forward by Zeno of Elea in the fifth century BC. According to the logic advocated by this paradox, *a watched arrow cannot fly*.

There is another aspect to this scenario, which we dub: "What's in the picture?" Until now we have considered a wide-view angle of the picture taken by EE, wherein Henry's position is recorded, so that the camera literally measures Henry's position. If instead a close-up picture is taken by EE, this is akin to a measurement with a binary outcome: is Henry there or not? As a result of this measurement he collapses to a state of being where she took the picture. There is yet another, counterintuitive, option. If EE takes a picture of her alterego SE, Henry will most probably not be seen there, since he has collapsed to being single, far from SE. Despite the fact that he is not in the picture, such a measurement discloses Henry's position and thereby changes his state. This effect is essentially *exclusion by measurement*, whereby EE *prevents* Henry from reaching the matrimonial state by repeatedly photographing the bride.

If Henry's initial position is unknown to EE, she cannot take a close-up of Henry in order to "freeze" him. However, if she repeatedly takes pictures of the bride she can still prevent them from uniting in a state of marriage, even though she does not know where Henry comes from. This is because no matter what his original state, the coherent transfer will have only a small probability, which will be reduced almost to zero by EE's pictures.

To summarize Henry's wedding dream, all quantum processes occur via coherent transfer of probability amplitudes from one quantum state to another. By repeatedly measuring the process at a much higher rate than the "natural" time-scale of this process (the transfer time in our story), one can essentially "freeze the evolution" and prevent any changes in the quantum state by the quantum Zeno effect (QZE). There is no classical analog to this amazing quantum phenomenon, whose implications will be further discussed in this chapter.

We next wish to extend the discussion to more general scenarios where the effects discussed above take a different character. Imagine that instead of coherent transfer towards SE, Henry would attempt to make a transfer to their matrimonial state through decay via his coupling to the environment, as in Chapter 9. EE could then still induce the QZE by interrupting this decay frequently enough by repeated measurements of Henry's state. The QZE condition would then amount to having such frequent interruptions that the *decay process would be indistinguishable from coherent transfer*! Therefore, we could replace the decay process—which involves (Chapter 9) energy transfer to many microscopic oscillator modes of the environment (the "bath")—by analogous energy transfer to a *single* oscillator mode, for the following reason. If the interruptions are highly frequent, there is not enough time between interruptions for different oscillator modes to get out of tune with each other (despite their different frequencies). Consequently, all modes keep evolving synchronously, and can thus be replaced by a single mode whose population oscillates as the sine-squared function (Figure 10.1).

Let us imagine that SE in our story is represented by the environment/bath. Then, the frequently interrupted process would involve transfer between only two states: 1) an excited (single) Henry along with an unexcited SE; 2) conversely, an unexcited (married) Henry and an excited (married) SE. In complete analogy to the previous analysis, EE could then exploit the QZE to strongly suppress the probability of their marriage by extremely frequent measurements. Yet, as explained in Section 10.2, if her measurements turn out not to be as frequent as those leading to the QZE, then the *opposite effect* can occur—the so-called *anti-Zeno effect (AZE)*—whereby the probability of Henry's marriage would *increase* instead of decreasing, as EE reduces the interval between measurements (the picture time). Both the QZE and the AZE have been experimentally observed beyond dispute, and both have diverse practical applications, as discussed subsequently.

10.2 THE QUANTUM ZENO AND ANTI-ZENO EFFECTS

An isolated quantum system prepared in an eigenstate of its Hamiltonian (an energy state) remains in that state indefinitely, even if this state is excited. Yet no system "is an island", to paraphrase John Donne; every system interacts and becomes entangled with its environment (a bath), as discussed in Chapter 9.

Therefore, an eigenstate of the system is not an eigenstate of the entangled system–bath complex, but rather a superposition of system–bath composite states, some of which are excited. This superposition state is unstable, because the system–bath interaction that couples its constituent states can cause transfer between these states. Upon averaging this superposition state over the (experimentally inaccessible) bath states, we find that almost any initial state of a quantum system is at least partly excited and thus unstable, so that it must decay to the lowest-energy ground state. This decay ends at the ground state provided the bath is empty of quanta (has zero temperature) and thus cannot re-excite the system (Chapter 9). The question we raise here is: what is the *time-dependence* of such decay?

E. Wigner (Chapter 3) and V. Weisskopf (Germany, later USA) were the first to address this question within QM in 1932. The decay process they considered was the de-excitation of a two-level system (TLS)—say, an atom—that is only weakly perturbed by its coupling with a bath—say, a continuum of electromagnetic/photonic or acoustic/phononic modes (Chapter 9). They found that the TLS excitation probability decays *exponentially in time*, to a good approximation, which turns out to be the same as the Markov memoryless approximation (Chapter 11).

Exponential decay is a process that occurs at a *constant rate*. In 1935, E. Fermi (Italy, later USA) introduced the *Golden Rule* (GR) for calculating this rate in the universal form:

$$\text{Decay Rate} = 2\,\pi \times \text{DOS}\,(\omega_0) \times (\text{coupling strength})^2$$

where the density of states (DOS) of the bath (the number of bath states in a unit volume) and the system–bath coupling strength are both evaluated at ω_0, $\hbar\omega_0$ being the TLS resonance energy (Chapter 2). This formula describes the rates of a vast variety of quantum decay processes, such as spontaneous emission of a photon from an atom, or phononic relaxation of a defect in a solid, to name only two.

Another key application of Fermi's GR is to the leakage of a quantum particle's wavefunction from a spatial region where it is confined by a potential barrier to the continuum of energy states that describe the particle's motion outside the barrier. This distinctly quantum-mechanical process, discovered by G. Gamow (USA, 1928) and termed *tunneling* (Chapter 13), occurs at a rate that conforms to Fermi's GR formula. An example of such a process is beta-decay of radioactive isotopes, whose lifetime is the inverse tunneling rate of an electron through the nuclear barrier.

For decades, the consensus had been that unstable quantum states decay exponentially according to the GR, at a nature-given rate that cannot be altered, until in 1958 L. Khalfin (USSR) proposed that at sufficiently short times, quantum decay may be non-exponential; i.e., the decay rate may be time dependent. However, he did not provide a clear recipe that could relate this time-dependence to the characteristics of the system–bath interaction.

Such a recipe, albeit crude, was provided by L. Fonda (Italy, 1972), who related the slowdown of the decay to the time it takes the emitted quantum (say, a photon) to traverse the system (say, an atom or a molecule). As long as the emitted quantum is within the system, the two are correlated, constituting one entity. Only after the emitted quantum has left the system is their correlation lost and genuine decay of the unstable state takes place. According to Fonda, the duration of non-exponential decay is roughly this *correlation time*.

Fonda introduced another notion into the discussion of decay: the *effect of frequent projective measurements* on the decaying state. Although projective measurements are mainly aimed at providing information on the quantum state, their other effect on the state is the interruption of its evolution by the randomization of its evolving phase. Fonda surmised that if the measurements repeatedly interrupt the evolution within the correlation time, the resulting decay will be slowed down. However, he had difficulty in suggesting processes where non-exponential decay could occur on non-negligible, experimentally detectable time scales, as discussed subsequently.

B. Misra and E. C. Sudarshan (USA, 1979) introduced an altogether different approach to the decay of unstable quantum systems. They discovered a universal rule for any quantum (unitary) evolution whereby *the energy spread of a closed system that is not prepared in an energy state grows as the time squared at sufficiently short times* (Section 10.4). According to this rule, upon interrupting the evolution by projective measurements sufficiently frequently, a striking effect is revealed which Misra and Sudarshan termed the *quantum Zeno effect* (QZE) (Sections 10.3 and 10.4) and considered to be *universal*: the more frequent the measurements, the slower the decay! More precisely, the QZE causes the decay rate to scale as the time interval between consecutive measurements. Thus, in the limit of vanishingly small intervals between measurements, which correspond to the system being continuously measured, its decay rate vanishes and the evolution freezes!

The conclusions of Misra and Sudarshan concerning the universality of the QZE and its ability (in principle) to freeze the evolution were widely accepted in the 1980s and 1990s, especially after the experimental demonstration of this effect by W. M. Itano et al. (USA, 1990). The platform for this demonstration was a three-level atom in which one field-driven transition was used as a frequently-measuring "meter" (E, for environment) of the evolution in another, slowly oscillating transition (S, for system) (Figure 10.2). The "meter" transition performed a measurement whenever its upper level spontaneously decayed via photon emission to the lowest level. The experiment indicated that as the "meter" action grows faster, the monitored oscillation turns into monotonic decay that becomes progressively slower (Figure 10.2). This unequivocal confirmation of the QZE consolidated the consensus regarding its *universality*.

However, there were dissenting voices too. A. S. Lane (UK, 1983) argued that frequent measurements could cause not only slowdown but also, under the right conditions, *speedup of the decay*. Yet a mechanism to explain the origin of and the conditions for these two opposing trends remained obscure. Such a mechanism and the ensuing conditions were put forward by A. Kofman and G. Kurizki (Israel, 1996, 2000, 2001), who took an approach that fundamentally differed from that of their predecessors: they treated the system and the bath distinctly, unlike Misra and Sudarshan and many others. Kofman and Kurizki's treatment has yielded the universal formula (Section 10.4) that generalizes Fermi's GR, in that it accounts for frequent interventions (not only measurements of the system, but also phase or amplitude changes in the system–bath coupling; Chapters 11 and 12).

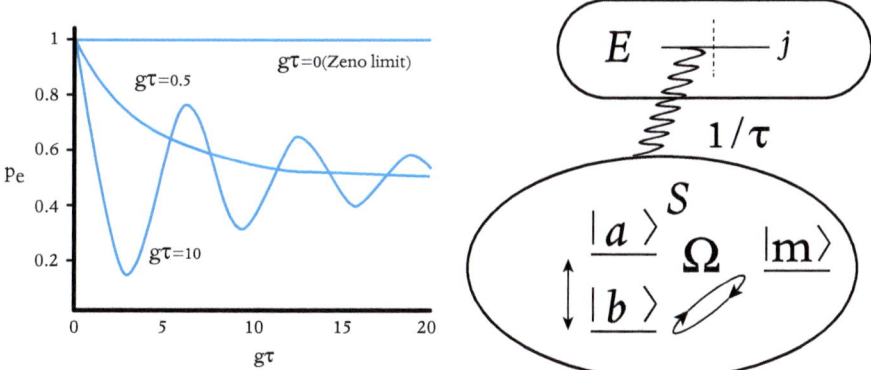

Fig. 10.2 (Left) The outcomes of the experiment by Itano et al. The nearly sinusoidal Rabi oscillation between $|b>$ and $|m>$ in S turns into nearly exponential decay of $|m>$ as the rate of measurements (monitoring) grows. The QZE arises when the monitoring rate $1/\tau$ becomes much higher than the Rabi oscillation rate Ω (when the interval between measurements, τ, is short enough). (Right) Schematic view of the QZE experimental setup by Itano et al. Three-level system coupled to an environment (see text).

The picture that underlies Kofman and Kurizki's analysis is that of a bath composed of infinitely many oscillator modes, each at a different frequency, that are coupled to a simple system, say a TLS. Over long times, an initial excitation of the TLS may be irreversibly transferred to the bath if there are bath modes at resonance with the TLS. By contrast, over much shorter time intervals the resonance condition is relaxed, and bath modes with very different frequencies may become excited, owing to the time–energy uncertainty relation (Chapter 6) that allows the randomly shifted TLS transition energy to match (i.e., be resonant with) bath modes at the same energy (Figure 10.3).

An extreme case occurs when the time interval between consecutive measurements is so short and so frequent that according to the time–energy uncertainty relation, the TLS energy spectrum is much broader than the energy range of the bath modes. In this case, the overlap between the broadened (random) TLS spectrum and the bath spectrum is suppressed, so that most bath modes do not match (are off-resonant with) the TLS energy and thus cannot be excited. This case is the *decay slowdown* regime; i.e., the QZE (Figure 10.4).

The opposite trend is obtained for much less frequent measurement rates, such that the overlap of the TLS and bath spectra is *augmented* by the broadening. This trend—termed by Kofman and Kurizki the *anti-Zeno effect* (AZE)—gives rise

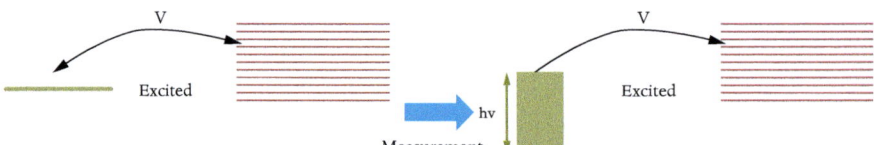

Fig. 10.3 Measurements broaden an excited level, analogous to phase randomization by collisions at a rate ν, drastically changing the level decay into the bath.

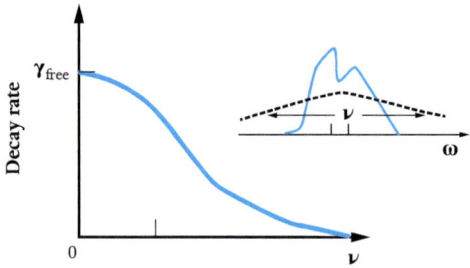

Fig. 10.4 Decay rate as a function of the measurement rate ν, under the QZE conditions, shown in the inset. The decay rate decreases with respect to the unmeasured (normal) rate γ_{free} as the measurement rate ν starts to exceed the bath spectral width.

to *enhanced decay rates* (decay acceleration). More precisely, the decay rate governed by the AZE grows as the *inverse* of the time interval between measurements $(1/\tau)$, whereas the QZE-governed rate grows as their time interval (τ). Thus, *the alleged universality of the QZE can be refuted* (compare Figure 10.4 and Figure 10.5 upper inset).

Remarkably, *the AZE is found to be more common (ubiquitous) than the QZE!* The reason for this is that the extremely frequent measurements required for the QZE (but not for the AZE) correspond to an energy spread of the TLS that may be prohibitively large. The higher energies in this spread may not comply with the physical process which the QZE is meant to suppress!

Few examples may elucidate these limitations on the occurrence of the QZE. One example concerns the radiative decay of an excited atom via spontaneous emission of radiation. The appropriate correlation time, according to Fonda's considerations, is then expected to be 10^{-18} sec, which is the atomic-electron orbital size (Chapter 1) divided by the speed of light. Yet the QZE requires much shorter time intervals, since the emission of radiation also involves high-energy (relativistic) motion of the electron that wiggles on much finer time-scales than the atomic-electron orbital period. It then follows from the time–energy uncertainty relation that the QZE may occur when the frequently-measuring

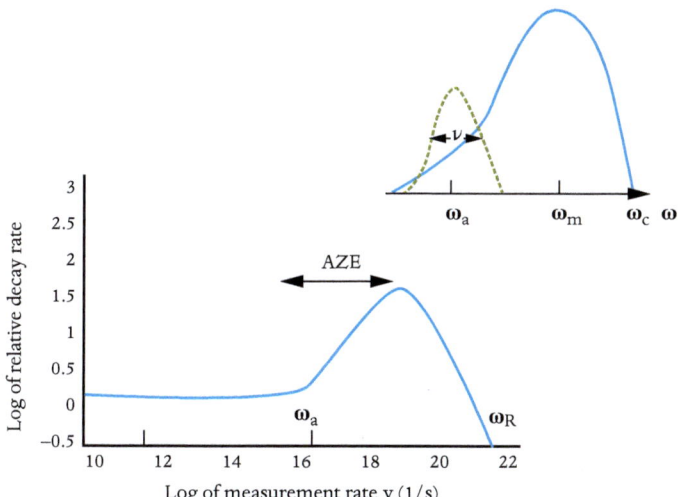

Fig. 10.5 AZE dependence of the decay rate. (Inset) Conditions for the AZE. (Graph) The dependence of the logarithm of the normalized decay rate $\log_{10}(\gamma/\gamma_{\text{free}})$ on $\log_{10}(\nu)$ for a spontaneously emitting hydrogenic state. The atomic transition frequency $w_a = 1.55 \times 10^{16} s^{-1}$, whereas the relativistic cutoff $w_R = 7.76 \times 10^{20} s^{-1}$, and Bohr frequency $w_B = 8.50 \times 10^{18} s^{-1}$. The AZE range is marked.

apparatus incurs an energy spread of nearly 1 million electron Volts (!) into the emitting electron. This energy spread is comparable to the entire energy of the electron according to Einstein's relativistic formula $E = mc^2$, where m is the electron mass and c is the speed of light. According to quantum electrodynamics, such an enormous energy spread allows for the creation of an electron–positron pair out of emptiness (the vacuum). Thus, when the emitting electron is perturbed by the frequently-measuring apparatus at a rate compatible with the QZE, it may, instead of slowing down its decay, end up in an altogether different process: pair creation (Figure 10.6). Similar pair-creation effects may occur when the beta-decay of an electron from the nucleus is probed on time-scales conducive to the QZE. By contrast, for the AZE to take place, the measuring apparatus needs to probe the system on *much slower time-scales* and hence incur a much smaller energy spread that does not jeopardize the outcome of the decay process. This explains why the AZE is more ubiquitous than the QZE.

Despite the inherent limitations on the QZE, it remains an extremely versatile tool of controlling decay. Especially useful is a generalized type of the QZE that can protect not only a single state but a whole manifold thereof (a subspace). Another, not less useful, range of applications is associated with the generalized AZE, which allows for augmented bath effects on a decaying state or subspace. Analogous QZE-like or AZE-like control methods for a system coupled to its environment are applicable to decoherence slowdown or speedup. These methods are the two complementary paradigms of decay and decoherence control that have been formulated by G. Gordon, A. G. Kofman and G. Kurizki (Israel,

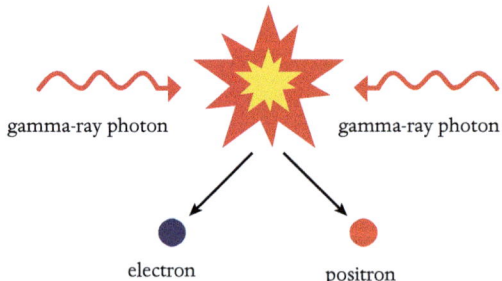

Fig. 10.6 QZE that slows down the decay of an excited electron in an atom requires measurement rates so high that they fall within the frequency range of gamma rays. Two gamma-ray quanta can be converted into a pair of electron (particle) and positron (antiparticle) by a quantum electrodynamic process called *pair creation*. Such a process is, however, incompatible with slowing down decay via the QZE, which fails in this case.

2000–08), and P. Facchi and S. Pascazio (Italy, 2002). In recent years the diversity of applications of these control paradigms has grown rapidly, as will be shown in Chapters 11 and 12.

10.3 IS TIME (OR CHANGE) AN ILLUSION?

The core of human experience is our *being enslaved by time*: it sets the limits to our existence, heals our injuries and collects the debt for our pleasures. Is there an escape from this slavery? One course taken to this end has been to challenge the reality of time. Both Western and Eastern philosophy and religion have, over millennia, been scornful of the randomness and inconsequence of temporal phenomena, as opposed to their reverence for eternal truths. A good example is Spinoza's standpoint that time is an utterly insignificant mode of reality that engenders unworthy sensations: fear, regret or hope. Therefore, according to Spinoza, our reason must transcend or ignore time. But is it possible? It is probable that very few humans have risen to the spiritual height requested by Spinoza.

Can we at least expect a clear statement on the reality of time as viewed by physics? Alas, it is fair to say that physics has not yet chosen between the opposing schools of thought on time that have existed since the early days of ancient Greek philosophy (fifth century BC): those of Heraclitus of Ephesus—who likened time to "a river in which we cannot bathe twice", since the fleeting moment is the only reality—and of Parmenides and his disciple Zeno of Elea, who utterly denied the reality of time and viewed any change as an illusion. Why is it so difficult for physics to decide which of these standpoints is true? Partly because the definition of time is elusive. The option we shall discuss here identifies the flow of time with change in the positions or properties of objects.

The standpoint of Heraclitus may be linked to the atomism of Democritus of Abdera (Chapter 7), who attributed temporal changes to erratic collisions between atoms. The modern counterpart of this atomistic theory is expressed by statistical physics, in both its classical and quantum versions, whereby large ensembles of atoms incessantly change via collisions, but conserve their average properties. The standpoint of Parmenides and Zeno, by contrast, was based on purely abstract, logical arguments. It relied on the maxim "whatever is, is, and whatever is not, is not". This maxim would only be contradicted by any *true change*, because such a change implies that something that has not existed starts to exist, or vice versa.

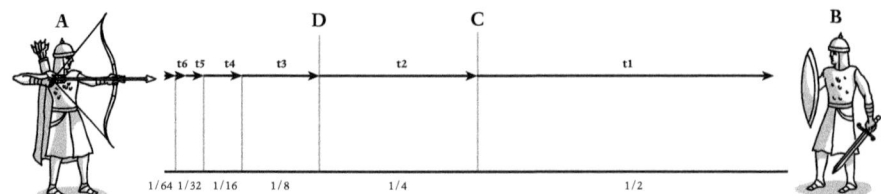

Fig. 10.7 Zeno's arrow paradox (see text).

Zeno strived to substantiate this maxim by paradoxes that contested the possibility of motion. The most poignant one is the arrow paradox: the more frequently we watch an arrow in flight, the closer its consecutive positions will be to each other, so that observations with null separation will pin the arrow to a single point in space where it must remain indefinitely! In more detail (Figure 10.7): to reach point B from point A, the arrow must pass through the midpoint C, and to reach C it must pass through D, which is halfway between A and C, and so on. Hence, the arrow must necessarily cross between two points that are so adjacent to each other that they coalesce to a single point. Thus, one is forced to conclude that *the arrow does not move at all*, or, colloquially, "A watched arrow does not fly." A paradox similar to Zeno's is that of Achilles, the fastest runner in Greece, and a turtle. If the turtle is given a head-start, then, by the same arguments as above, Achilles will never catch up with it.

These paradoxes were formally refuted following the introduction of the integral and differential calculus by Newton and Leibnitz in the seventeenth century, and the mathematical "limit" by Weierstrass in the nineteenth century. These concepts imply that the division of an integral into an infinite number of segments may still yield a finite result; in the paradoxes discussed previously, this result is a finite non-zero velocity, which is the derivative of distance with respect to time. Conversely, the sum of infinitesimal intervals which the arrow traverses (or the sum of infinitesimally converging distances between Achilles and the turtle) is an integral that corresponds to a finite distance, even though each interval is of zero extent. Thus, after all, concluded the proponents of calculus. the arrow would fly and Achilles would overtake the turtle.

This rigorous refutation should have put an end to Zeno's paradoxes concerning the impossibility of change. Yet they persisted in some circles. Jerome K. Jerome, in his *Three Men in a Boat*, stated that "a watched pot does not boil." Borges—literator and erudite (Chapter 4)—was captivated by the beauty of Zeno's paradoxes.

Strikingly, forty years after Borges, B. Misra and E. C. G. Sudarshan (USA) revived Zeno's standpoint in the form of the quantum Zeno effect (QZE), whereby change can be nearly "frozen out" by sufficiently frequent observations, and this prediction has been confirmed by experiment (Section 10.2). Does the QZE imply that the arrow can be pinned down in mid-air if it is closely watched? Here we dwell on the fundamentals of the QZE that have been elucidated by the theoretical work of A. Kofman and G. Kurizki (Israel) and P. Facchi and S. Pascazio (Italy), and experimentally confirmed by M. Raizen (USA). These works (Section 10.2) have underscored the need to view an observed quantum system as being coupled to its environment: essentially, *any system is an open quantum system* (Chapter 9). It then follows that the quantum-mechanical way of describing consecutive observations, as intermittent coupling of the system to a meter, necessarily perturbs the natural decay of the state of the system to the environment, because the system–meter coupling and the system–environment coupling compete (do not commute) with each other. The outcome of this competition is a random spread of the energy of the system states by an amount that grows with the rate of observations. This energy spread is a result of the time–energy uncertainty relation, the relevant time being the interval between observations. As the energy spread becomes so large that the states of the system go completely out of tune (resonance) with the states of the environment (Figure 10.4), the QZE sets in, and the decay of the state of the system is inhibited. Equivalently, this is the limit in which the system–meter coupling varies so strongly that the system has no time to interact with the environment, and the effect of the environment is averaged out in time. In a sense, this is a confirmation of Zeno's argument that *change* (caused by the environment) *is an illusion*, as it is up to the observer to prevent it by appropriate observation.

However, the QZE comes at a price which was unknown to Zeno or to any physicist prior to the advent of QM: observations of *a system coupled to its environment* cost energy, since they must separate the system from the environment; i.e., decouple the two. This energy cost grows with the energy spread of the system states which is enforced by the time–energy uncertainty principle. Hence, the complete realization of the Zeno paradox—which corresponds to infinitely frequent observations and thus to an infinite energy spread—is unphysical. Such a spread cannot be accommodated by any physical system, and would require an observer to have infinite energy resources.

For many natural decay processes—such as radiative decay of excited atoms in open space or nuclear radioactive decay—the energy cost of a train of pulses corresponding to intermittent system–meter interactions would be too high for a QZE with present technologies. Yet before long we may witness demonstrations of the ability to slow down such decay processes which had been viewed as immutable until the theoretical predictions mentioned previously.

The QZE and its derivatives have opened up promising avenues of controlling system–environment interactions, which we shall survey in Chapters 11 and 12. But their principle moral is related to *the active role that an observer inevitably plays in QM*. We have seen (Chapter 4) that this role is active because the observer must choose which observable is to be measured—a choice that may profoundly affect the results. Now we see that this role extends to the energy resources that the observer must invest according to the rate of observations. In a loose paraphrase on the Red Queen's advice to Alice (in *Alice Through the Looking Glass*) we may conclude that if you wish to remain still (i.e., not evolve) you have to be probed (observed) as hard as you can. But this conclusion does not contradict that of Parmenides, that *change (or its absence) is up to us humans, acting as observers, hence it is subjective (illusory) rather than objective* (Figure 10.8).

Fig. 10.8 The Zeno and anti-Zeno effects as viewed by Zeno and by William Tell watching an arrow in mid-air.

Zeno's Arrow

Achilles, mighty athlete, is so vain:
He gives the turtle head-start, as a bait.
But lo! Their distance measured, time and again,
Dooms them to an eternal stalemate.
The mystery of space and time unfolds
As arrows are launched, but would not fly,
Only because someone their course beholds,
And nature hastens with the observer to comply.

10.4 APPENDIX: SLOWING DOWN THE EVOLUTION

This appendix is divided into two parts: (i) a simple explanation of the quantum Zeno effect (QZE) for coherent transfer (described in Henry's wedding dream at the start of Section 10.1); and (ii) the QZE for decay processes (described in the last part of Section 10.1).

10.4.1 QZE for coherent transfer

As described in Section 9.4, the Rabi button generates coherent transfer between two states. In our story, if we denote Henry's single (bachelor) state as an arrow pointing up and his married state as an arrow pointing down, then the entire state representing Henry's transfer towards SE has the form:

$$|\psi(\mathbf{t})> = a_1 \uparrow (t)|\uparrow> + a_1 \downarrow (t)|\downarrow> = \cos(\Omega t)|\uparrow> + \sin(\Omega t)|\downarrow>$$

where the probability amplitudes of these two states oscillate as a cosine or sine, respectively. Now consider EE taking a picture after a short time τ, when $\Omega\tau \ll 1$, since then EE's paparazzi snapshots can stop Henry's approach towards SE, as shown below. For such a short duration, $\sin\Omega\tau \approx \Omega\tau$. Hence, Henry's probability of collapsing next to SE when EE takes his picture is then:

$$p_{marriage} = (\Omega\tau)^2$$

After every unsuccessful attempt, Henry's chances for matrimonial happiness are reduced to zero, so that he must start again "from scratch". The total duration of the complete uninterrupted transfer is $T = \pi/2\Omega$, while Eve may interrupt the transfer T/τ times, if she keeps the interval between pictures fixed. The overall (*accumulated*) marriage probability during the ceremony then becomes

$$p_{marriage} = (\Omega\tau)^2 \times \pi / (2\Omega\tau) = \Omega\tau\pi/2 \propto \tau$$

where the first factor on the right-hand side is the success probability in each attempt, interrupted after a time-interval τ, and the second factor is the number of interruptions during the transfer. The last step on the right-hand side indicates that the result is *proportional to the time interval* between pictures. This result means that the more frequently the pictures are taken—the shorter the interval between interruptions—the smaller Henry's *accumulated* probability of getting married.

Any quantum process involving coherent transfer between an initial and a final state by means of a constant coupling (which determines the Rabi frequency) is described by sine and cosine oscillations, as shown above. Hence their short-time quadratic behavior is generic, and allows for a QZE if interrupted by frequent measurements. Indeed, experimenters use this effect for their benefit whenever they wish to halt or at least slow down the dynamics generated by unwarranted forces, as discussed subsequently.

10.4.2 QZE for a decay process

While Henry's dream dealt with coherent transfer down the aisle, quantum systems commonly encounter decay or decoherence processes due to their interaction with the environment (bath). As described in Chapter 9, decay may occur due to the coupling of the quantum system to a plethora of excitable tiny detectors (Eve's sensors in that story) which are typically identified with the microscopic/nanoscopic oscillator modes of the bath. Let us revisit the relevant equations:

$$\frac{da_{\uparrow 0}(t)}{dt} = -R(t)a_{\uparrow 0}(t)$$

$$R(t) = \int_0^t dt' e^{i\omega_a(t-t')} \sum_j \hbar^{-2}\mu_j^2 e^{-i\omega_j(t-t')}$$

Here the first equation describes the change in the probability amplitude of the initial excited state of the system, corresponding to the 0th unexcited state of the environment (bath). The time-dependent rate of change of the initial-state probability amplitude, $R(t)$, is determined by an integral over two factors: the first is the oscillating exponential of the transition energy between the excited and ground states of the system, and the second is a sum over all bath modes involving the oscillating exponentials of the bath-mode frequencies weighted by the corresponding system–bath coupling strengths squared. In the Zeno regime—when measurements or other types of interruption are separated by extremely short time intervals—t is much smaller than the oscillations of the exponential phases associated with either the system energy levels or the bath mode frequencies, so that $\omega_a t \ll 1$, $\omega_j t \ll 1$. In this regime, one can approximate the oscillating exponentials as follows: $e^{i\omega_a(t-t')} \approx 1 + i\omega_a(t-t')$, $e^{-i\omega_j(t-t')} \approx 1 - i\omega_j(t-t')$. Since t is extremely small, t^2 is even smaller and can be neglected. Hence, when computing $R(t)$, the terms within the integral satisfy $1 \gg \omega t \gg (\omega t)^2$, which means that we can indeed take into account only the terms that are either 1 or linear in t in each exponential function. This results in the simple decay rate:

$$R(t) = \sum_j \hbar^{-2} \mu_j^2 t = \Gamma^2 t \qquad \Gamma = \frac{1}{\hbar}\sqrt{\sum_j \mu_j^2}$$

Here, Γ is the instantaneous decay rate that is determined by the squared strength of the coupling to the bath. Upon plugging this approximate expression into the differential equation for the excited-state probability amplitude, we find:

$$\frac{da_{\uparrow 0}(t)}{dt} = -\Gamma^2 t a_{\uparrow 0}(t)$$

This equation has the solution: $a_{\uparrow 0}(t) = e^{-\frac{\Gamma^2 t^2}{2}} a_{\uparrow 0}(0)$. The assumption of extremely short times, $\Gamma t \ll 1$, enables us to approximate the exponential function again as follows:

$$a_{\uparrow 0}(t) = \left(1 - \frac{\Gamma^2 t^2}{2}\right) a_{\uparrow 0}(0).$$

This expression has the *same quadratic form* as its coherent-transfer counterpart, where we have similarly approximated the cosine and sine terms. This proves that *frequently interrupted decay is indeed indistinguishable from similarly interrupted coherent transfer,* as we argued in Section 10.1. Hence, the same conclusions regarding the QZE for coherent transfer are valid for its counterpart for decay; namely, the probability of the system to remain in its initial excited state is extremely close to 1 if it is subject to sufficiently frequent measurements or other interruptions.

Yet how can it happen that decay, which results from the coupling of a system to a multitude of modes (channels) of the environment (bath) at various frequencies, behaves similarly to coherent transfer, which involves a *single* coupling channel? A close examination of the system–bath interaction proves the pictorial argument advanced in Section 10.1. During sufficiently short time intervals, all of the bath modes *oscillate in phase* despite their diverse frequencies, and so all of them exhibit quadratic time-dependence within the same time interval. This means that the combined system–bath complex *behaves coherently* during this very

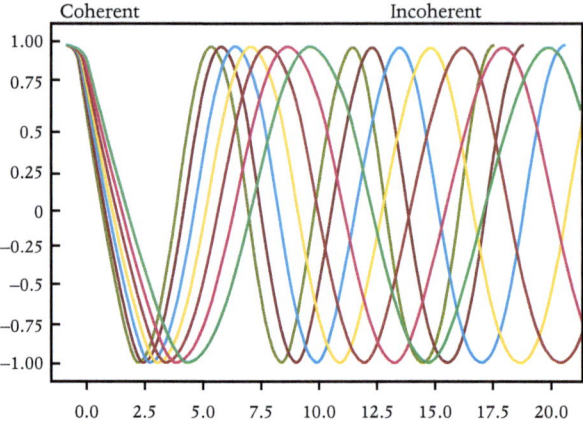

Fig. 10.9 Amplitudes of many oscillator modes with different oscillation frequency (indicated by color) as a function of time. At short times (less than 2.5 in this case) all oscillation frequencies follow a similar evolution pattern, so that their amplitudes are synchronized or, equivalently, phase-coherent. As time goes on, the phases of oscillators with different frequencies deviate more and more from each other; i.e., their oscillations become incoherent.

short stage (Figure 10.9). Hence, the same arguments that lead to the "freezing" of the state evolution in a simple system, where coherent transfer occurs, apply to the combined system–bath complex, showing that the QZE is common to both decay and coherent transfer. Only as soon as the various bath modes start oscillating differently (out of phase) and thus acquire a variance in phase, which occurs at the stage when $\left(\omega_j - \omega\right) t \approx 1$, can the bath no longer be treated as a single coherent entity. In fact, the various bath modes then tend to oscillate *out of phase with each other*, so that the interruption of the decay at this stage may aggravate the deviation of the system–bath complex from coherence, and thereby cause *faster-than-normal decay*. This trend, which is even more ubiquitous than the QZE, is the anti-Zeno effect (AZE) (Section 10.2).

The fire is a hot, big quantum bath.

The fire interacts with me. The interaction entangles the fire with me.

If I'm measured, this entanglement is gone. Its energy is released as heat.

At the right rate of measurements, the fire will take up all the heat and I will cool down.

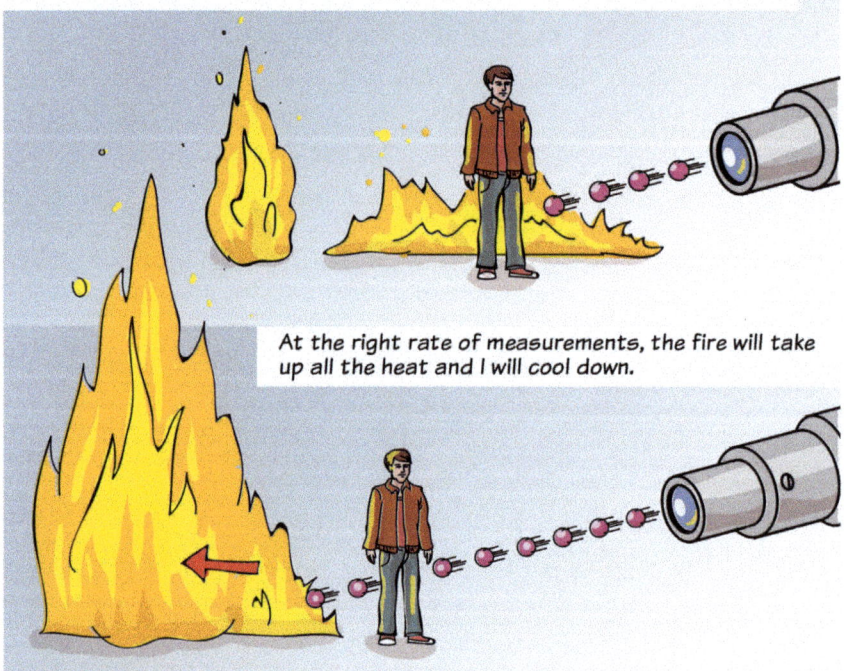

CHAPTER 11

Can Quantum Measurements Control Temperature?

11.1 COOLING DOWN A HEATED RELATIONSHIP

Henry is eager to shape his precious quantum crystal into a lens that can endow light with unprecedented quantum properties (as we shall learn in Chapter 14). To this end, he decides to solicit the help of Prof Raman, a quantum physicist who has become a spiritual guide but still practices quantum-optical and quantum-material technologies, in which he has unique ("holistic") knowhow. As Henry gets to Prof Raman's Center for Spiritual and Scientific Studies and starts discussing with him the shaping of his quantum crystal, he finds out to his dismay that Eve has already been there, and, having learnt of Henry's new ideas concerning the quantum crystal, has just left the place enraged. In her fury, Eve has slammed the door so hard that one of the chandeliers at the shrine has tipped over and set the place ablaze. (Or did she do more than just slam the door? We cannot tell.) Henry manages to lead Prof Raman to the fire escape stairs just before they collapse. Prof Raman thus manages to get to safety, but Henry remains trapped in the burning building. Can QM effects rescue Henry from the fire?

To answer this question we must first revisit the relevant notions of heat and temperature in the quantum world (Chapter 9). From Henry's adventures in Pisa with his quantum rocket (Chapter 6), we know that quantum states can correspond to different energy levels, so that one must spend energy in order to jump from a lower to a higher energy level, as Henry did with his rocket. A quantum system with a given temperature is in a statistical "mixture" of states corresponding to different energy levels with fixed probabilities. At high temperature the system has large probability to occupy the higher energy levels, whereas at low temperature it has an appreciable probability only to reside in the lowest energy levels. The fire at Prof. Raman's Center can be described in QM language as a high-temperature environment consisting of many tiny molecules that populate high energy levels with large probability.

How can Henry be burnt by such "quantum fire"? Henry is a quantum system at the lowest temperature possible, since his quantum state corresponds to the lowest energy level. When the fire is about to consume him, the environment starts interacting with him, thereby causing transfer from the lower to the higher state. The initial state of the combined system is separable into hot-environment and cold-Henry states. As the interaction progresses, an entangled, joint, state of Henry and the environment emerges, wherein the states of the hot molecules become entangled with Henry's quantum states. This increasingly entangled state is ominous for Henry. Due to energy conservation between the initial and the final states of the joint system, Henry's high-energy quantum states become increasingly more occupied by receiving energy from the environment molecules. Before long, the physiological functioning and even the structure of Henry's molecules will be disrupted beyond repair. Can anything be done, with the help of QM, to avoid this calamity?

Henry is at a loss. In his despair, a strange thought crosses his mind: only Eve can help him out! As a last resort he sends her a message: "m'aidez, SOS, m'aidez!" He is relieved to see her come to his rescue in the nick of time.

The entanglement process described here is, as seen before (Chapter 9), a coherent transfer of excitations between the initial and final quantum states. In this episode, the initial quantum states are separable whereas the final quantum states are entangled. As discussed in Chapter 10, such transfer is described by oscillations of the probability amplitudes that build up until they reach their largest value and then recede back to zero. This oscillatory nature of the process will prove to be crucial for Henry's survival.

Henry sends Eve another message: "QND". Being a competent quantum physicist, she understands that he begs her to measure his energy in a fashion known as quantum non-demolition (QND) measurement, which does not change or demolish Henry's energy state. Nor does it affect the state of the environment (the fire), on which it does not act. Thus, Henry's energy state is to remain exactly as before the measurement: perilously hot. Yet, as Henry and Eve know, the QND measurement *destroys the entanglement* between the environment and the system. While Henry remains in a hot quantum state, he is no longer entangled with the fire. Eve performs the requested QND measurement, but they both realize that he is still in danger, as the entangling process between the fire and Henry starts anew after the measurement, bringing about an increase in Henry's temperature once again!

Another message from Henry is quick to follow: "QZE". As in Henry's wedding dream (Section 10.1), Henry and Eve have in mind highly frequent measurements which they expect will protect Henry from further heating. The reasoning is as follows. Since the burning process is a coherent exchange or transfer of probability amplitudes between the initial and final quantum states discussed previously, the quantum Zeno effect (QZE) can come to the rescue. By measuring Henry at a sufficiently high rate, the hope is to be able to *freeze* his state (pun intended).

Eve attempts an extremely rapid succession—a volley—of QND measurements in order to achieve the QZE, but to her and Henry's dismay he becomes *even more heated* than before the measurements! Thus, highly frequent QND measurements result in what may be termed *Zeno heating*.

Is Henry then doomed to be consumed by the fire, or is there still hope for another quantum rescue? Since there is no time to lose, they hectically grope for a solution, until . . . Eureka! They both recall that the heating of Henry's quantum states oscillates: sometimes Henry is hotter, and sometimes he is *momentarily* cooler, despite the overall increase in his temperature by the end of the oscillation. Hence it is possible to adapt the interval between measurements to the time instants at which Henry is cooled down! What Eve has to do is reduce the rate of QND measurements so as to measure Henry at those "cool" instants (pun intended again). She shoots a slower volley of QND measurements, and then asks Henry: "Well?" The answer comes: "Hurrah! I'm cooler now!." After many such volleys of measurements at carefully selected intervals, Eve notices that Henry starts shivering in the midst of the raging flame! It is time to stop and let him out, Eve decides. A tender reconciliation follows Henry's jump out of the throngs of fire into Eve's arms . . .

The quantum dynamics that has enabled Eve's incredible feat requires closer examination. Immediately after being disentangled by the first QND measurement, the joint state of Henry and the environment starts entangling again, causing Henry's heat-up. Then, however, the tide turns. The heat flow between the environment and Henry is reversed: The environment starts heating up, whereas Henry starts cooling down. The oscillatory nature of energy transfer that underlies their entanglement on very short time-scales has been the cause of Henry's cooling, as Eve has managed to measure Henry by QND measurements, at just the right times, when he is instantaneously cooler than before. Since the QND measurement does not alter Henry's energy state, he will remain cooler

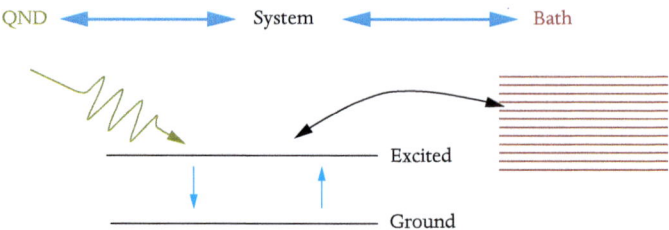

Fig. 11.1 Exchange of energy between the two-level system (TLS) and the hot bath (red) is disturbed by QND measurements of the system energy (green).

until an oscillatory entangling process with the environment resumes. If Eve repeats the measurements at the right rate, each such interaction with the fire will actually *cool Henry*. This effect is the opposite of Zeno heating and is thus termed *anti-Zeno cooling*.

Eve's counterintuitive quantum rescue process is only partly similar to that of decay prevention in Section 10.1. The similarity stems from the QM time–energy uncertainty relation (Chapter 6): the shorter the interval between Eve's measurements, the broader the energy spread of the system (Henry's) states, and therefore the broader the range of environment states that are resonant with Henry's. Thus, when Eve performs excessively frequent measurements on Henry, the higher energy states of the environment can dump their energy into Henry's, causing his overheating. In order to achieve chilling instead of heating, the timing of QND measurements by Eve must be correctly chosen (Figure 11.1). If the measurements are performed at the right moments during the oscillatory exchange of energy between Henry and the fire, when he is momentarily in his coldest state, the result is an increasingly cooler Henry.

This counterintuitive effect—anti-Zeno cooling—has been confirmed by experiment (Section 11.2). In our story, not only Henry has been cooled off by this effect, but also his fiery rapport with Eve.

11.2 QUANTUM ZENO HEATING AND ANTI-ZENO COOLING

As discussed in Section 10.2, the decay of quantum system excitations into a zero-temperature bath can be either slowed down or sped up if perturbed or interrupted by frequent measurements in the QZE or AZE regime, respectively. N. Erez, G. Gordon and G. Kurizki (Israel, 2008) predicted a strange variation on this theme by analyzing what happens if one attempts to repeatedly observe or measure the temperature of a system—say a TLS—that is immersed in a thermal bath. The analysis showed that when measurements are extremely frequent, such that the QZE takes place (i.e., the relaxation is slowed down), then, much to the observer's surprise, the temperature will continue to increase from one measurement to the next; i.e., the system will progressively heat up. Equally surprising is the outcome of less frequent measurements, such that the AZE—i.e., relaxation speedup—occurs. The temperature will then continue to decrease with the number of measurements; i.e., the system will progressively cool down (Figure 11.2).

To grasp the anomaly of these effects—namely, their apparent incompatibility with standard thermodynamic rules—we shall revisit the concepts of equilibrium thermodynamics. Let us specifically consider an ensemble of TLS at equilibrium with a bath, characterized by their temperature T. The meaning of this temperature is understood by comparing the excitation (upward) transition

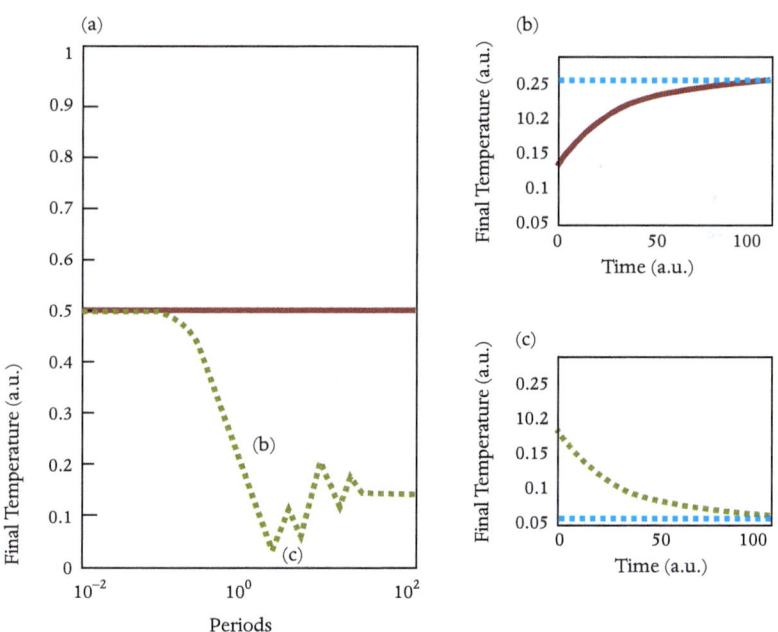

Fig. 11.2 TLS temperature as a function of measurement rate. (a) Final temperature attained as a function of the time interval (period) between measurements. The temperature initially decreases, then increases (b), then decreases again (c), and continues to oscillate until the long-time limit is reached. (b) Temperature evolution as a function of time under QZE heating. (c) The same, under AZE cooling. Here the arbitrary units (a.u.) correspond to the TLS energy (for temperature) and its inverse (for time).

rate at which quanta from the bath are absorbed by the lower level of the TLS to the exponentially smaller de-excitation (downward) rate at which quanta are emitted by the upper level. The ratio of these two rates is known as the Boltzmann factor of the equilibrium ensemble. This factor is determined by the TLS energy divided by the temperature T (in energy units). If the TLS energy is much larger than T—say ten times larger—then the Boltzmann factor is negligible (exponentially small), and so is the population of the upper level of the TLS. This is the low-T limit wherein only the lower level is appreciably populated. By contrast, the high-T limit is obtained when the TLS energy and T are nearly equal, as are the two-level populations—the Boltzmann factor being nearly 1 (Section 11.4).

What can be expected as regards the equilibrium? According to the First Law of Thermodynamics, the temperature of the TLS ensemble must remain constant as long as its contact with the bath persists. According to the Second Law of Thermodynamics, any disturbance of the TLS–bath equilibrium state must increase the total entropy. Under the standard thermodynamic assumptions that the bath entropy does not change, while the TLS–bath coupling is negligible, any disturbance must *increase* the TLS entropy; i.e., *heating up of the TLS* is expected!

None of the above expectations holds true in the scenario considered by Erez, Gordon and Kurizki (EGK)! As a result of measuring the TLS temperature (which is the same as measuring its energy), the TLS initially heats up and later cools down, then heats up again, resulting in an oscillation of its entropy or heat flow. This behavior is in apparent contradiction with the First and Second Laws of Thermodynamics as stated previously. In fact, what is violated are not these laws but the standard assumption that the TLS–bath coupling is negligible! Albeit small, their coupling has (often disregarded) consequences—primarily that they are *entangled* at equilibrium (Chapter 9). Hence, by observing and measuring the energy of the TLS, one affects the joint, non-separable state of the TLS and the bath, and thereby breaks their equilibrium and initiates their evolution. This leads to alternating heating and cooling of the TLS. The price for this effect is that the observer must invest at least as much energy in the TLS measurement as the mean TLS–bath coupling energy, thereby effectively decoupling them during the measurement.

Thus, the unexpected post-measurement departure of the TLS from equilibrium requires unconventional thermodynamic considerations that account for system–bath entanglement changes caused by measurement. The surprising result of the EGK analysis that the entropy oscillates depending on the rate of measurements stems from a related effect: violation of the standard thermodynamic assumption that the bath state does not change, even though it is very large and only weakly coupled to the TLS. Changes in their mean coupling energy cause the entropy and heat flow to oscillate between the TLS and the bath, so that each of them alternates between heating up and cooling down, at different time intervals, partly at the expense of energy invested by the observer.

A closer examination of the TLS heating and cooling by frequent measurements relates these effects to the broadening of the TLS levels discussed in Section 10.2. This broadening can change the population ratio of the excited and ground levels in the TLS because of the combination of two effects which play out differently in the QZE and AZE regimes. (i) In the QZE regime both the absorption and emission rates of the TLS are reduced (slowed down), as detailed in Section 10.2. The two rates become nearly equal when the levels are so extremely broadened that they become indistinguishable (Figure 11.3). Then, since at low temperature T the population is almost entirely in the lower state, absorption transfers more population to the excited state than the other way around, which is why the TLS is heated up. (ii) By contrast, in the AZE regime, the rates of emission and absorption are both enhanced. However, since the level broadening by frequent measurements is then much smaller than in the QZE regime, the absorption rate in the low-T limit is exponentially smaller than the emission rate, so that more population is transferred downward than upward, thus leading to measurement-induced TLS cooling! These EGK predictions, which bring out uniquely quantum aspects of thermodynamics, have charted out a new frontier termed *thermodynamics under observation*, or *measurement-controlled thermodynamics*, of coupled quantum systems.

The EGK predictions were confirmed by an experiment performed by G. Alvarez, D. D. B Rao, L. Frydman and G. Kurizki (Israel, 2010). In this experiment, the TLS was a spin-½ carbon nucleus, and the bath consisted of three protons which were likewise spin-½ TLS (Figure 11.4). Normally, there

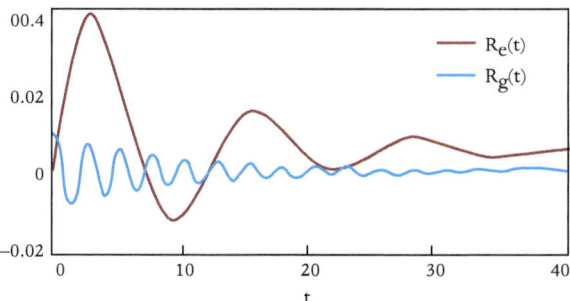

Fig. 11.3 Absorption (R_e) and emission (R_g) rates as a function of time (in units of the inverse TLS frequency) for a TLS coupled to a bath of forty oscillator modes at different frequencies (see text).

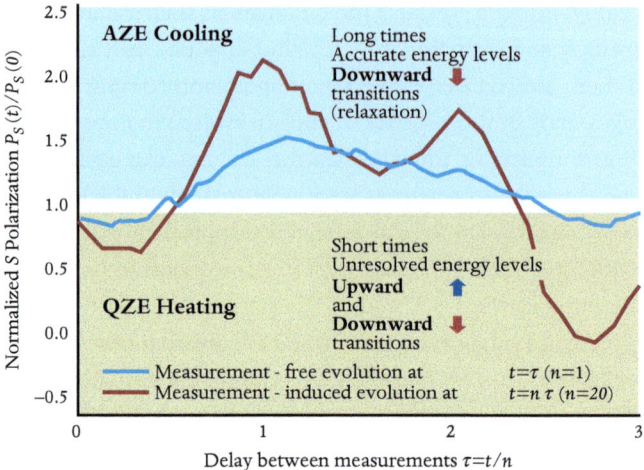

Fig. 11.4 System (proton-spin) coupled to a bath (three carbon spins) under repeated measurements: heating (spin-depolarization) under QZE conditions; cooling (spin-polarization increase) under AZE conditions. Results of the experiment by G. Alvarez, D. D. B. Rao, L.Frydman and G. Kurizki (2010).

is no excitation exchange between the two subsystems because their spins are off-resonant with each other. However, when the carbon nucleus is subject to frequent measurements of its spin energy, then toggling of the excitation between the carbon and the protons takes place, resulting in either cumulative heating up or cooling down of each subsystem, depending on the rate of measurements. The implications of this effect are greater than its curiosity: it allows us to heat up or cool down nuclear spins of live tissue by acting at the appropriate rate on their neighboring spin-probe, even though it is off-resonant with the spins in the tissue. Such spin-selective and highly localized cooling of nuclear spins in the tissue may dramatically improve magnetic resonance imaging (MRI) sensitivity in the years to come.

11.3 FIDDLING WITH THE ARROW OF TIME: ANOMALOUS THERMODYNAMICS

The Second Law of Thermodynamics is probably the most universal law of physics. It has an astoundingly broad realm of applications, ranging from subatomic to cosmological scales. It has been hailed by Einstein and Feynman, among others, as the most enduring natural law. Not less remarkable is the ability of the Second Law to withstand the transition from classical to quantum mechanics *without change,* because the quantum–classical transition mainly concerns the *microscopic* description of reality, and not the thermodynamic or statistically averaged *macroscopic* picture. Thus, according to the Second Law, Von Neumann's entropy of an open quantum system in contact with a thermal bath *increases* (or remains unchanged), just like its classical counterpart (the Boltzmann–Gibbs entropy). This expresses the growth in the lack of information or the degree of disorder of the system (Chapter 9). Hence, the time directionality is as well defined by the Second Law for open quantum systems as for their classical analogs.

Engraved in stone as the Second Law and the ensuing time directionality may be, they too rest on certain assumptions: the arrow of time as an expression of the Second Law is conditioned on *coarse graining in time* (finite time interval averaging) which may fail. The crucial assumption underlying coarse graining (Chapter 9) is that we do not attempt to examine the system at each instant, but rather allow some time between consecutive observations. This enables us to ignore fast-oscillating superpositions of eigenstates or their observables and follow the slower trend of the system to reach thermal equilibrium with the bath; i.e., attain the bath temperature. The Second Law is synonymous with this equilibration trend that does not hold, as neither does coarse graining, for open systems in the domains of validity of the QZE and the AZE (Section 10.2).

We saw an example of such failure in Section 11.2, where we attempted to repeatedly observe the energy of a TLS embedded in a thermal bath by coupling the TLS to an energy meter. In that scenario, consecutive observations separated by a sufficiently long time always yield the same mean energy, corresponding to the bath temperature, but if we repeat the observations at intervals comparable to the inverse TLS energy, then its mean energy alternates between higher (hotter) and lower (colder) values as the interval is varied, as does the TLS entropy (state purity), irrespective of the bath temperature. Such alternation signifies the wiggling of the arrow of time rather than its strict directionality, manifested by steady (monotonic) entropy increase of the TLS according to the Second Law, if the bath is assumed not to change. Such anomalous behavior was shown to be a consequence of the time–energy uncertainty relation: the energy spread of the TLS becomes too large on such short time intervals to comply with the equilibration trend of thermodynamics. The rapid coupling and decoupling of the TLS and the meter dominates the dynamics and overwhelms the competing effect of the TLS–bath coupling. Essentially, if the TLS energy spread exceeds the energy difference between the levels, it becomes unclear as to which level (prior to the measurements) was the "upper" or the "lower" one. Contrary to the normal thermodynamic assumption, quantum emission can take the TLS upward (from the "lower" to the "upper") state, causing its heating, and vice versa. Quantum absorption can take the TLS downward, causing its cooling.

The foregoing anomalies of temperature and entropy on ultrashort time-scales probed by repeated measurements can be summarized as *hot is cold and cold is hot*, to paraphrase the witches in Macbeth. One may wonder: is it necessary to resort to frequent measurements and the dynamics they induce to reveal these anomalous deviations from equilibrium? Alas, QM leaves us no choice but to specify the observation procedure the way we did, aiming for minimal intrusion (non-demolition). The anomalous dynamics is an unavoidable price for trying to single out an observable of the system and ignore its environment on short time-scales. In terms of information, relaxation proceeds by information leakage from the system to the environment, but frequent monitoring causes the information to flow back and forth between the system, the environment and the meter. This is the reason why the arrow of time loses its meaning under such conditions.

It is intriguing to speculate over the role of anomalies associated with the QZE and the AZE in the *primordial universe*. The arrow of time expresses the entropy growth in the universe since the Big Bang, but what might have preceded it? Was there, perhaps, a stage where the mean time and energy wiggled and produced the Big Bang as we know it? We may never know the answer to such a surmise. Be it as it may, the preceding discussion might have perhaps convinced Parmenides even further that his standpoint (Section 10.3) was right: without the observer who can control not only the pace but even the direction of time flow, there is no point to define time at all. It is entirely a figment of our imagination or thought.

Panta Rey ("Everything flows")

Everything flows, Heraclitus wrote,
What's now here is never to return.
But our science is challenging this quote:
At small scales, everything can backward turn!
Imagine how life will change if we can find
Ways to revert the lethal time flow,
When happy flames, extinguished in our mind,
Will be rekindled in their old, warm glow.

11.4 APPENDIX: DISRUPTING THE HEATING PROCESS

In this appendix we elaborate on the system–environment interaction and the quantum dynamics induced by its change, as follows. 1) We revisit their interaction Hamiltonian (Section 9.4) and consider its extraordinary implications on energy conservation in quantum processes. 2) We then examine the mean interaction energy that corresponds to the Henry–fire coupling rather than to either Henry or the fire, and show that this energy is *negative*. 3) We next focus on Henry's quantum state, represented by his density matrix. 4) We then introduce the quantum non-demolition (QND) measurement, which has properties that relate to previously discussed aspects of distinguishability and visibility (Chapter 8), and show that this measurement eliminates the Henry–fire interaction energy. 5) Finally, we dwell on Henry's dynamics *after* the measurement and show that it can indeed cool him.

1) In Section 9.4 we introduced the interaction Hamiltonian between Henry's and Eve's sensors. Here, the environment represents the fire and can start from the high-energy states $|1>_j$, and not only the lower energy states, as before. The Hamiltonian in Section 9.4 had only two *energy conserving* terms, corresponding to either increasing the environment energy while decreasing Henry's, or vice versa. In a fuller description there are two more terms in the interaction Hamiltonian that defy classical energy conservation, but not quantum energy conservation. The entire system–environment interaction is

$$H_{int} = \sum_j \mu_j (\, |\downarrow><\uparrow| \, ||0>_j < 1| + |\uparrow><\downarrow| \, ||1>_j < 0| + |\downarrow><\uparrow| \, ||1>_j < 0|$$
$$+ |\uparrow><\downarrow| \, ||0>_j < 1|)$$

The two last terms appear to defy energy conservation, since they either decrease the energies of both the system and environment (third term) or increase both energies (last term). How can both Henry and the fire gain energy at the same time? Where does this energy come from? In quantum mechanics, the conservation of energy applies only to *expectation values* or averages. As discussed in Chapter 6 in the context of the time–energy uncertainty relation, energy is also a quantum operator. While *only* the last term truly defies energy conservation, the third term, which is mandatory for preserving the unitarity of the Hamiltonian, makes up for it. Hence, together, on average, they conserve energy.

2) The most general Henry–fire state is given by:

$$|\psi(t)> = a_{\uparrow0}(t)e^{i\omega_a t} |\uparrow>|0> + a_{\downarrow0}(t) |\downarrow>|0>$$
$$+ \sum_j a_{\uparrow j}(t)e^{i(\omega_a+\omega_j)t} |\uparrow>|1>_j + a_{\downarrow j}(t)e^{i\omega_j t} |\downarrow>|1>_j$$

Recalling how energy is measured (Chapter 6), we can calculate the *interaction energy* by taking the expectation value of the interaction Hamiltonian:

$$E_{int} = <\psi | H_{int} | \psi> = \sum_j \mu_j \left(a_{\downarrow0}a_{\uparrow j}e^{i(\omega_a+\omega_j)t} + a_{\downarrow0}a_{\uparrow j}e^{-i(\omega_a+\omega_j)t} \right.$$
$$\left. + a_{\downarrow j}a_{\uparrow0}e^{-i(\omega_a-\omega_j)t} + a_{\downarrow j}a_{\uparrow0}e^{i(\omega_a-\omega_j)t} \right)$$

Since $e^{ix} + e^{-ix} = 2\cos(x)$, careful consideration of the first two terms combined and the last two terms combined yields the interaction energy:

$$E_{int} = <\psi \mid H_{int} \mid \psi \geq 2 \sum_j \mu_j \left(a_{\downarrow 0} a_{\uparrow j} \cos\left(\left(w_a + w_j\right) t\right)\right.$$

$$+ a_{\downarrow j} a_{\uparrow 0} \cos\left(\left(w_a - w_j\right) t\right)$$

This energy can have *negative values* at specific times, depending on the oscillation frequencies.

3) We next wish to find Henry's density matrix. To this end, we use the *trace out* operator introduced in Chapter 8 to trace out the fire. This gives us Henry's state in a clearer manner:

$$\rho_H = <0|\psi><\psi|0> + \sum_j <1|\psi><\psi|1>_j$$

Henry's state thus has four contributions:

$$\rho_H(t) = \rho_{\uparrow\uparrow}(t) \, |\uparrow><\uparrow| + \rho_{\downarrow\uparrow}(t) \, |\downarrow><\uparrow| + \rho_{\uparrow\downarrow}(t) \, |\uparrow><\downarrow| + \rho_{\downarrow\downarrow}(t) \, |\downarrow><\downarrow|$$

$$\rho_{\uparrow\uparrow}(t) = |a_{\uparrow 0}(t)|^2 + \sum_j |a_{\uparrow 1}(t)|^2$$

$$\rho_{\downarrow\downarrow}(t) = |a_{\downarrow 0}(t)|^2 + \sum_j |a_{\downarrow 1}(t)|^2$$

$$\rho_{\downarrow\uparrow} = \rho_{\uparrow\downarrow}^* = a_{\downarrow 0} a_{\uparrow 0} e^{-iw_a t} + e^{-iw_a t} \sum_{j,j'} a_{\downarrow j} a_{\uparrow j'} e^{-i\left(w_j - w_{j'}\right)t}$$

As can be seen, the diagonal terms of the density matrix, $\rho_{\uparrow\uparrow}$ and $\rho_{\downarrow\downarrow}$, are positive and represent the *probabilities* of Henry being hot and cold, respectively. Henry's off-diagonal terms are those that contribute to the interaction energy: they arise from the interaction Hamiltonian that has only off-diagonal, $|\uparrow><\downarrow|$ and $|\downarrow><\uparrow|$, terms.

4) We next examine Eve's rescue endeavor performed by a quantum non-demolition (QND) measurement, whose principles were introduced in Chapter 8. This measurement consists in the creation of a fully entangled system (Henry-)detector state and then averaging it over the detector states. In our case, all it tells us is that when Henry is in the $|\uparrow>$ state the detector is in the detected state $|d>$, and when he is in the $|\downarrow>$ state the detector is in the undetected state $|u>$. Hence the combined system–detector density matrix is a simple extension of Henry's density matrix:

$$\rho_{H+D} = \rho_{\uparrow\uparrow} \mid d>|\uparrow><\uparrow|<d\mid +\rho_{\downarrow\uparrow} \mid u>|\downarrow><\uparrow|<d\mid +\rho_{\uparrow\downarrow} \mid d>|\uparrow><\downarrow|$$
$$<u\mid +\rho_{\downarrow\downarrow} \mid u>|\downarrow><\downarrow|<u\mid .$$

Averaging over the detector states is *tracing the detector out*. Since this procedure is completely ignorant of the final state of the detector, it is known as *non-selective measurement*. However, an extraordinary effect arises when the detector is traced out in the combined density matrix:

$$Tr_D \rho_{H+D} = <d|\rho_{H+D}|d> + <u|\rho_{H+D}|u> = \rho_{\uparrow\uparrow}|\uparrow><\uparrow| + \rho_{\downarrow\downarrow}|\downarrow><\downarrow|.$$

Here, the *off-diagonal terms of the combined density matrix vanish*, and nothing else happens! Since the diagonal terms determine Henry's temperature, the non-selective measurement *by itself* does not help Henry to cool down. It does, though, eliminate his interaction energy with the fire, which is only determined by the off-diagonal terms and now *vanishes*, regardless of its earlier value. Most importantly, it resets the Henry–fire entanglement to zero. Henry is no longer coupled to the fire immediately after the (brief) measurement, though the fire is still there! The interaction then starts anew.

5) How does this QND measurement save Henry from the quantum fire? The final piece of the puzzle lies in the *post-measurement dynamics*. Henry's diagonal terms obey very similar rules to Henry's decay down the mine. Skipping the details, the rates of change of Henry's upper-state and lower-state populations have the form:

$$\frac{d\rho_{\uparrow\uparrow}(t)}{dt} = -\frac{d\rho_{\downarrow\downarrow}(t)}{dt} = R_\downarrow(t)\rho_{\downarrow\downarrow}(t) - R_\uparrow(t)\rho_{\uparrow\uparrow}(t).$$

We observe that there is total population conservation, so that the rates of change of the two populations balance each other. As for Henry's decay in Chapter 8, the $-R_\uparrow(t)\rho_{\uparrow\uparrow}(t)$ term is the rate at which Henry's upper state decays to his lower state in the mine: in the burning building it represents the rate at which Henry cools down. However, there is another term in this dynamical equation, $R_\downarrow(t)\rho_{\downarrow\downarrow}(t)$, which represents Henry's rate of heating up. The coefficients of the heating-up and cooling-down rates are:

$$R_\uparrow(t) = Re \int_0^t dt' e^{i\omega_a(t-t')} \sum_j \hbar^{-2}\mu_j^2 e^{-i\omega_j(t-t')}$$
$$R_\downarrow(t) = Re \int_0^t dt' e^{i\omega_a(t-t')} \sum_j \hbar^{-2}\mu_j^2 e^{i\omega_j(t-t')}$$

Here the symbol *Re* in front of the integral means "real part", since the coefficients of the *probability* rates of change cannot be complex, in contrast to the *probability amplitude* rates. While both rates appear to be similar, they differ by the *sign of the environmental oscillating exponent*. As a result, the cooling rate is determined by the fact that the exponent is $\omega_a - \omega_j$ instead of $\omega_a + \omega_j$ for the heating rate.

Let us delve deeper into the *post-measurement dynamics* described by the previous equations. At a very early time after the measurement, as long as the quantum Zeno effect (QZE) prevails, the rate coefficient grows linearly with time, $R(t) \sim t$, with no dependence on ω_a or ω_j. Hence, during that time, we have $R_\downarrow(t) = R_\uparrow(t) = R_{QZE}$, meaning that Henry is cooling down at the same speed as he is heating up! Since there is more of cool Henry than hot Henry (as long as Henry is not totally burnt, $\rho_{\downarrow\downarrow}(t) > \rho_{\uparrow\uparrow}(t)$; i.e., Henry has *finite temperature*) we then find:

$$R_\downarrow(t)\rho_{\downarrow\downarrow}(t) - R_\uparrow(t)\rho_{\uparrow\uparrow}(t) = R_{QZE}\left(\rho_{\downarrow\downarrow}(t) - \rho_{\uparrow\uparrow}(t)\right) > 0$$

This means that Henry is becoming hotter despite (or because of) the Zeno effect!

However, there is a time interval after the measurement when the oscillations of the rate coefficients $R(t)$ not only affect their magnitude but can even render them *negative*. If the right time interval is chosen between measurements, these quantum coherent oscillations of $R(t)$ lead to what would be classically impossible, as Eve found out:

$$R_\downarrow(t) < 0, R_\uparrow(t) < 0$$

$$R_\downarrow(t)\rho_{\downarrow\downarrow}(t) - R_\uparrow(t)\rho_{\uparrow\uparrow}(t) < 0$$

During this time interval, the energy flows *in the "wrong" direction*, contrary to what is stipulated by thermodynamics—from the cooler Henry to the hotter fire! This phenomenal effect can only be achieved due to the quantum coherence between Henry and the fire, after they are rattled by the QND measurement.

To summarize the rather complex rescue scenario, we have seen that Henry and the fire are initially entangled, with an interaction energy that varies with time. By using a QND measurement, Eve manages to temporarily decouple Henry from the fire by eliminating the interaction energy. This, in itself, does not cool Henry, but it reboots the interaction or resets its time. In the initial stage thereafter, Henry and the fire again coherently oscillate due to their renewed interaction. Choosing the right moment, when Henry's oscillatory energy is near its deepest trough (Figures 11.2 and 11.4), the energy or heat flow between Henry and the fire is temporarily reversed, resulting in Henry's cooling down at the expense of the fire heating up, instead of what is normally expected. Repeating these measurements results in cooling Henry to a safe temperature.

CHAPTER 12

Can Dephasing be Controlled?

12.1 HENRY CONTROLS HIS DEPHASED DESCENT

Some time after Henry and Eve have decided to suspend their rivalry, Eve flies with Henry on her airplane above a spectacular but ominous rainforest. They are on a perilous mission to transform their quantum crystals into amplifiers, flying above guerilla-controlled territory. They try to evade radar detection, but then. . . a sudden blast (of a shell or a missile?) cripples the airplane! Alas, it turns out that the crystal was not powerful enough to counter the blast! Henry and Eve bail out, but while Eve has her parachute strapped on and can safely descend, Henry finds himself in peril, because his quantum suit, as a result of its high-power discharge, has uncontrollably split Henry into four quantum versions, only one of which is able to get hold of the parachute. In order to survive, all his versions must recombine exactly where the parachute is, or else hit the ground separately.

Henry, being a skilled quantum physicist and having been through many quantum adventures, quickly considers his options. He can ask Eve to measure him and thus collapse him to a single place. However, he knows that there is no way to tell *a priori* which state he will collapse to. Since there is a 75% chance that he will collapse to the wrong place, far from the parachute, and face a dismal ending, Henry decides not to take chances. He seeks a *deterministic rather than a probabilistic* procedure to ensure with 100% certainty that he will emerge as a full, classical Henry in the right place.

He recalls his escape from Eve's stakeout with his motorcycle (Chapter 3), where he used constructive and destructive interferences to deterministically control his final location. This appears to Henry to be his best option, and so he presses the Recombine button expecting the four versions to coincide and form the fully classical Henry near the parachute. But alas, after the Recombine operation, he splits again into four quantum versions of himself!

Why has the Recombine button not worked? As in Chapter 3, Henry has counted on destructive interference to eliminate his unwanted versions and on constructive interference to make him appear his classical self again near the parachute. Now these effects have failed him and left him one last chance before perdition.

Henry knows that in order for interference to work properly, the relative phases of all his quantum versions must be fully in control. However, Henry's ejection from the airplane has scrambled his phases beyond recognition, since some versions have apparently gathered more phase than others. Having no control over the phases, he cannot deterministically guide one version to become classical and others to disappear via destructive interference.

What Henry experiences is a form of decoherence termed *dephasing*, which is extremely common in quantum systems. Its underlying mechanism can always be traced back to entanglement of the system with its environment. Yet what matters is that they *appear to be random*, and this *randomization* of the relative phases between superposed quantum states prevents their controlled interference. Is there no hope for Henry?

In his mounting state of panic, Henry suddenly hears Eve shout: "Your phases are linear!" He then recalls an almost forgotten insight into *dephasing* that sparks a flicker of hope in his mind: perhaps he can control the phases despite their apparent randomness so as to successfully recombine! Although he does not know the phase of each state, he suspects, as does Eve, that the phases have grown *linearly* with time, which is typical for dephasing processes. The rest of the long-forgotten lesson on dephasing then comes to his mind in a flash (there is nothing like a noose round your neck to sharpen your faculties, they say): by flipping the phases of his quantum versions—i.e., just changing the phases' *signs*—all his quantum states will revert to their original phases and the accumulated random phases will cancel out, if he waits as long as it has taken them to dephase. Henry quickly turns his phase dial in all his quantum versions to the *opposite sign*, and waits for the exact time that has elapsed from their initial dephasing, and then . . . success! All his quantum versions have reverted to the same relative phase! They recombine into a classical Henry with 100% probability at the place where he can strap on his parachute, just in the nick of time.

Henry's hair-raising adventure has taught us a remarkable lesson. If a system is subjected to dephasing, such that each of its quantum states accumulates a random phase linearly with time, one can perform a quantum operation that reverses all the phases and thereby *undo* the random dephasing, even without knowing what those phases were! However, the phases return to their original values *only at a specific moment of time*, which is twice the duration of the initial dephasing period (Figure 12.1). If one does not seize the moment, dephasing will keep growing linearly with time. Hence, in order to effectively combat this decoherence effect, one must repeatedly perform the phase-flip operation. This periodic repetition—which must be frequent enough to prevent the random phases from becoming non-linear in time—is reminiscent of the quantum Zeno effect (QZE) in Chapters 10 and 11; but, contrary to the QZE which invokes measurements that *destroy coherence*, the periodic phase-flipping operations are *coherent*, and hence guarantee that the relative phase will be zero (as long as the process is linear with time).

Since a great many quantum systems are subject to decoherence in the form of dephasing from one source or another, the technique described here for combatting such effects is ubiquitous. Henry owes his life to this technique, as it has allowed him to catch his parachute just in time. This technique is indispensable in nuclear-spin resonance measurements—particularly in magnetic resonance imaging (MRI), so that many MRI patients are indebted to it, as explained in Section 12.2. One of its popular names is *bang-bang*, inspired by the impulsive, repeated nature of the operations. Its original name was *spin echo*, as the dynamics after the phase-flipping echoes the dynamics before this operation.

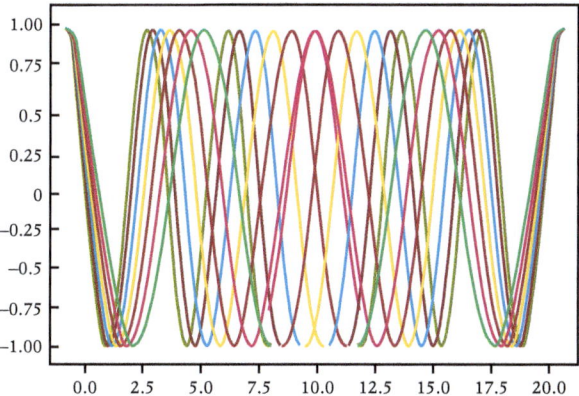

Fig. 12.1 The phases of many oscillators with random frequency differences grow linearly in time (in units of the red oscillator quarter-period) and thus dephase. At $t = 10$ all the phases are reversed, and at $t = 20$ they all revert to their (same) initial value, thus undoing the dephasing.

12.2 DECOHERENCE AND ITS CONTROL

The violent shakeup of Henry during his ejection from the airplane has scrambled the phases of his superposition state, but similar processes whereby the phases of superposition states become corrupted occur naturally and unavoidably in any quantum system of interest, such as a qubit. If we follow its state in time, then sooner or later its interaction with its surrounding (environment) will cause the loss of its quantum phase coherence; namely, decoherence.

The notion of decoherence was introduced by J. Von Neumann (Chapters 3, 4 and 9) in his groundbreaking book on the foundations of QM in 1932. This notion followed from his discussion of quantum-measurement effects (Chapter 5): an ideal measurement projects a quantum superposition state onto an eigenstate, thereby destroying the phase coherence that existed among the eigenstates in the superposition; this process constitutes decoherence in its strongest sense. It occurs since the system and the measuring apparatus become entangled for the measurement to take place and then the apparatus is ignored, rendering the quantum state of the system mixed, or impure, without phase coherence.

Yet the notion of decoherence has its limitations. Its dynamical description (by R. Zwanzig, USA, and R. Kubo, Japan, in the 1950s and 1960s) has shown that it arises gradually rather than abruptly, over a time-scale called the *memory time of the environment*. Within this time-scale, which may be short, the effect of decoherence can be "undone", since the system and the environment can be disentangled by appropriate unitary operations. Such operations are the essence of the dynamical control of decoherence, to be discussed subsequently.

Finally, for most practical purposes decoherence need not be described by entanglement with the environment, which may be a highly exacting task, but may instead be attributed to the simpler effects of random noise induced by the environment. Noise is any process whose effect can only be described statistically over an ensemble of systems, but is unknown for any individual system.

Let us dwell on the latter mode of description, which underlies the tale in this chapter. The notion that decoherence is caused by noise was put forward by F. Bloch and I. Rabi (Chapter 3) with the advent of nuclear magnetic resonance (NMR) spectroscopy. They attributed to noise the rapid decay of the coherent signal obtained from a large sample of nuclear spins subject to a pulsed magnetic field. Such a field causes the precession of the spins (Section 10.2) about the axis of the static magnetic field that aligns the spins. Were this process coherent, the precession would be regular, yielding a signal that is periodic in time. Instead, the signal tends to decay over a time which Bloch and Rabi called $T2$. They identified two possible mechanisms for this decay. One is the spatially inhomogeneous character of the sample: individual spins are exposed to different magnetic fields, and therefore their precession occurs at different, unknown rates, thereby accumulating effectively random phases with time. This process is thus associated with static noise. Another source of noise, referred to as "dynamical", are the fluctuations of the field in time, causing the magnetically aligned spins to wobble and jitter in an unpredictable manner, thus giving rise to temporal phase randomness. Either one source of noise or the other may dominate.

For years, the decoherence caused by $T2$ processes (termed *dephasing*) was considered immutable. Yet in 1950 E. Hahn (USA) demonstrated the ability to overcome dephasing caused by static noise by a simple, ingenious method he called *spin echo*, for which he was later awarded the Nobel Prize. By this method, the phases of all spins that have been randomized over time interval T are reversed by a short pulse, so that after an equal time interval T they all return to their initial value and thereby cancel the dephasing. Their evolution in the second interval echoes that in the first interval. This is the method Henry Bar implemented in our tale. In 1954, H. Carr and E. M. Purcell (USA) and S. Meiboom and D. Gill (Israel) extended it to fight dynamical noise through more frequent application of echo pulses.

Since then, more sophisticated echo pulse sequences have been introduced to counter dephasing in NMR spectroscopy. Similar methods have been implemented for the protection of coherence between electronic states of atoms, based on the optical analog of the spin echo discovered by S. R. Hartmann (USA) in 1964. These methods are the basis of more recent applications (since the late 1990s) involving coherence protection from environment effects in quantum information processing and communication (QIPC) protocols implemented by qubits, such as teleportation (Chapter 14), since these protocols are highly susceptible to decoherence. Yet the very harsh demands of QIPC applications, requiring long coherence times, call for more advanced concepts of dynamical noise control than echo methods.

This survey may leave the readers with the impression that the understanding of decoherence and its control are well at hand; yet there are still unresolved conceptual issues to reckon with.

As demonstrated in Chapter 3, the result of an interference-based quantum operation is dependent on the relative phase between eigenstates in the quantum superposition. In some cases, any small change in phase can drastically alter the results of the experiment. But the truly disastrous situation is when this phase change is effectively random for each realization of the superposition, because the system is subject to noise. The result is called dephasing. As in the case of Eve's attack on Henry Bar's coherence, it may completely destroy the functionality of interference-based quantum operations.

In what follows, we provide some physical insights into dephasing. We then simply explain the technique known as spin echo, which is intended to fight dephasing from two generic noise sources: slowly fluctuating fields, and inhomogeneous spin ensembles.

The former type of dephasing is encountered, e.g., when the spin-qubit is realized by magnetic fields that split the energy eigenvalues associated with its two eigenstates. If the qubit is realized by magnetically sensitive nuclear spin states or electronic spin states in an atom, a magnetic field causes the levels of the spins that are aligned or counteraligned with the field to shift in opposite directions. An unpredictable change (fluctuation) in the noisy magnetic field causes a corresponding change in the level-energy separation and thus a change in the relative phase of the corresponding eigenstates. This change of relative phase may accumulate with time. If the phase fluctuations are slow, so that during a single experiment (realization) the phase remains constant but changes between experiments, this dephasing can be countered by spin-echo techniques, as will be described.

The other type of dephasing is typically encountered in setups that probe spins embedded in solids or liquids. It arises from the spatially inhomogeneous nature of the sample that is comprised of many spin systems. In these setups, many realizations do not require repeating the experiment many times, but rather conducting a single experiment on many systems at once. One then probes the density matrix of the statistical ensemble consisting of these systems, and obtains a signal corresponding to an ensemble average in a single experimental run. These systems are rarely identical, and due to differences in their local environment they may have different energy-level separations. Hence, in a single experiment, each spin system constitutes a realization that acquires a different, *a priori* unknown phase, again fulfilling the conditions for application of spin-echo techniques.

In both situations the individual spin system remains quantum-coherent during each realization. If the phase were to repeat itself for all realizations, it could be countered by measuring the resultant state, determining the phase and acting on the system to reverse its effects. Yet, since each realization encounters a different, effectively random phase, repeating the experiment would not enable the extraction of useful phase information.

To understand the essence of the spin-echo technique, consider that dephasing is associated with static noise acting over time t. The spin echo is then effected by a short pulse of the magnetic field that changes all phases by 180°. A useful analogy exists between this evolution and runners on a racetrack. At the first blow of the whistle they all start together, trying to keep their coherence (synchronism), but this synchronism is progressively lost due to their pace differences. At the second blow of the whistle, when the foremost runner is halfway through the race, all runners turn around so that the first runner is now last. After an equal time interval, all runners will finish the race together, as expected from the echo method (see cartoon).

12.3 QUANTUM CONTROL OF LIFE AND DEATH: IS CHANGE AN ILLUSION?

The main effort that has been guiding the development and implementation of decoherence control in recent years has been directed towards the protection of quantum coherence, entanglement and quantum information (QI); namely, their storage, processing and communication in systems intended to serve in quantum technologies. These emerging technologies include quantum computing, quantum cryptography and quantum teleportation, discussed in detail in Chapters 14 and 15.

This effort is a truly formidable endeavor, since these quantum technologies—particularly quantum computing—require extremely high fidelity on the one hand, and complex systems for their implementation on the other—both requirements leading to enormous sensitivity to decoherence (Chapters 9 and 12).

Challenging as this endeavor may be, perhaps there is a much further-reaching potential goal for decoherence control: can we someday use it to influence metabolism and even the process of dying? Classical thermodynamics does not allow the reversal of metabolic processes, since they are *irreversible* at every elementary step. But do such processes strictly abide by classical thermodynamics on all time-scales? In recent years, extensive (mainly theoretical) studies are being conducted in order to identify quantum effects in the simplest metabolic process: photosynthesis. These studies are motivated by G. Fleming's (USA, 2002) still unsettled findings that energy transfer within photosynthesis reaction centers may be quantum-coherent. If such a process or related processes are indeed found to exhibit quantumness, then we may start looking for their appropriate quantum control. As shown in Chapter 11, the Second Law of Thermodynamics and the ensuing irreversibility of energy transfer break down on extremely short time-scales. Hence the question is: can we devise quantum control that would stretch these time-scales so that they may influence metabolic processes? There is no fundamental reason why this should not be so!

The following directions of quantum control are likely to be pursued for controlling metabolic processes:

i) Dynamical QZE-like control (Chapters 10 and 12)

Dynamical QZE-like control is developing at a pace that may allow the slowdown of many-body decoherence in progressively more complex systems, provided they are coupled to thermal baths or noise sources with finite memory time, so that they can give rise to non-Markovian dynamics. It is encouraging that the rate at which such control must be applied does not grow much with the number of degrees of freedom of the system: as found by G. Gordon and G. Kurizki (Israel, 2011), the growth in the control rate is much slower than all other rates. Once we identify the quantum mechanisms that underlie metabolism, it might become possible to design and implement dynamical control that may prevent the death of living organisms over greatly extended time intervals compared to their natural lifetimes. This would vindicate Zeno's assertion that all change, including death, is an illusion.

ii) Protection of quantum coherence and entanglement by decoherence-free subspaces

A promising direction of control hinges upon the fact that biologically active molecular complexes, such as those responsible for metabolic processes, are composed of identical molecular units. Dynamical control of such processes can take advantage of the *indistinguishability* of interaction amplitudes between the bath and many molecular units because of their coupling to the common bath. This indistinguishability is the key to either constructive or destructive interferences; i.e., addition or subtraction of the amplitudes (Chapters 7–9), depending on the location of the individual units.

This phenomenon, first discovered by R. Dicke (USA, 1956), was later studied by S. Haroche (France, 1983) and M. O. Scully (USA, 2006) for short-time, transient emission of quanta by many identical atoms or molecules (Figure 12.2). These processes are manifested by *superradiance*—collectively enhanced spontaneous emission via constructive interference—or *subradiance*—the inverse process of collectively suppressed emission via destructive interference. Recent works by G. Gordon, W. Niedenzu and G. Kurizki (Israel, 2006, 2018) have shown that such collective enhancement or suppression can occur *continuously at all times* and be dynamically controlled in such a way that certain subspaces (sets of states) of the many atoms or molecules are *decoupled* from the bath

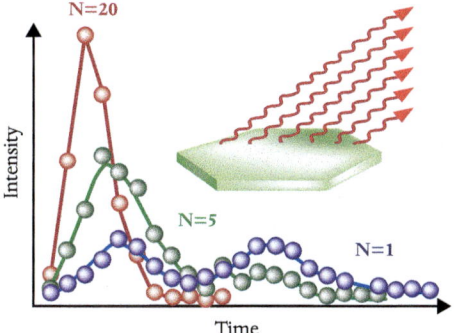

Fig. 12.2 Superradiance. (Inset) N identical atoms in an ordered (crystalline) array spontaneously radiate in phase; i.e., collectively, at an N-fold enhanced rate, termed *Dicke superradiance*. (Main panel) Superradiant intensity as a function of time for N atoms shows that the more atoms, the faster the radiation occurs (the intensity peak is shifted to shorter times).

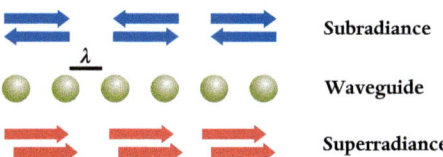

Fig. 12.3 Subradiance and superradiance controlled by initial preparation of the N-atom state in an ordered array. Red and blue are not the colors of the light; they are only used to distinguish between constructive (red) and destructive (blue) interference.

due to their destructive interference (Figure 12.3). These decoupled subspaces—termed *decoherence-free subspaces* (DFS) or *dark* subspaces—have been identified by P. Zanardi (Italy, 2002) and D. Lidar (USA, 2004) as being robust against the corruption of quantum entanglement or quantum information.

The employment of DFS for dynamical control may open the road to the preservation of quantumness in large, biologically active molecular complexes, and thereby to a much broader scope of quantum-controlled biological processes. However, the applicability domain of DFS is far more general, and is in fact *universal*. The reason is that a collection of identical systems invariably has a certain symmetry, so that when any two of these systems are exchanged, their interaction amplitudes with a common bath either preserve or flip their sign, corresponding to constructive or destructive interference, respectively—the latter case being that of a DFS. This universality of DFS may explain why certain interactions of multiparticle systems are suppressed or masked by means of protective, destructive interference among its constituents. Nature may be wearing an "invisibility cloak"—an idea put forward by Heraclitus of Ephesus, who, *c*.450 BC, dedicated to Isis (Artemis), the goddess of nature, the saying: "Nature likes to hide itself." This saying still poignantly summarizes the difficulty of revealing nature's secrets.

12.4 APPENDIX: BANG-BANG AS DEPHASING CONTROL

In Section 12.1, Henry's peril was caused by decoherence effects of the type known as dephasing: the incurrence of *random phases* into the time evolution of quantum states. Here we have in mind "classically random" phases—phases that are unknown to us but are *knowable* in principle, if one keeps track of the forces acting on each degree of freedom of the system. By contrast, the *principally unknowable* "quantum random" phases arise from the collapse of the system state (wavefunction) by its entanglement to the environment followed by the trace-out of the environment (Chapters 9–11). Here we present a simple mathematical explanation of dephasing and its control by the bang-bang or spin-echo method.

Let us examine an extremely simple Henry—one that goes left or right, as denoted by his two possible states: $|\leftarrow>, |\rightarrow>$. Initially his state is an equal superposition of the two: $|\psi> = \sqrt{(1/2)} \left(|\leftarrow> + |\rightarrow>\right)$. Henry wishes to recombine, preferably where the parachute is. In the absence of decoherence, a simple application of the Recombine button would suffice. However, as Henry drops rapidly towards the ground, one of his versions acquires a *random phase*, δ, with each passing second. Hence, his state evolves as:

$$|\psi> = \sqrt{(1/2)}\left(|\leftarrow> + e^{i\delta t}|\rightarrow>\right)$$

Henry is now in a serious jam. If he knew δ he could pick the time $\delta t = \pi$, which would result in a precisely known state, which, by using the Recombine button, could be reverted to a classical (whole) Henry again. Even better, by choosing the appropriate time he could control the state to which he would recombine, as in his interference act in Chapter 3.

Since δ is unknown to Henry, he uses the important notion that only the *relative phase* matters in quantum mechanics. He understands that flipping his states will be akin to reversing the accumulated phase. Let us see how this is done.

If Henry flips his states, then his state becomes:

$$|\psi> = \sqrt{(1/2)}\left(|\rightarrow> + e^{i\delta t}|\leftarrow>\right) = \sqrt{(1/2)}\left(e^{i\delta t}|\leftarrow> + |\rightarrow>\right)$$
$$= \sqrt{(1/2)}\, e^{i\delta t}\left(|\leftarrow> + e^{-i\delta t}|\rightarrow>\right)$$

Since only *relative phase* matters, $e^{i\delta t}$ outside the parenthesis has no physical meaning and can be removed, resulting in his flipped state:

$$| \psi(t) >= \sqrt{(1/2)} \left(|\leftarrow> + e^{-i\delta t} |\rightarrow>\right)$$

Henry continues to descend and accumulate the same random phase-rate, δ. This means that after an additional time t, the state $| \rightarrow>$ is multiplied by the factor $e^{i\delta t}$ that completely cancels out the *negative* phase factor that Henry has accrued over the preceding time interval t:

$$| \psi(2t) > = \sqrt{(1/2)} \left(| \leftarrow> + e^{-i\delta t} e^{i\delta t}| \rightarrow>\right)$$
$$= \sqrt{(1/2)} \left(| \leftarrow> + | \rightarrow>\right)$$

The amazing feat is that *regardless of the value of δ, the phase that has been accumulated disappears*. This stands in stark contrast to Henry's previous endeavors in which he needed full knowledge of his phase in order to control his quantum state. In the present scenario Henry has no clue what his phase is, but he knows how to effectively *reverse time* via flipping his states and changing the sign of the accumulated phase.

Random dephasing is associated with the notion of a quantum *ensemble*: many *identical replicas* of systems in different quantum states labeled by index j. The ensemble is characterized by the quantum density matrix, which describes how all the systems are measured together, and has the form:

$$\rho = \sum_j | \psi>_j < \psi | .$$

If many *dephased* systems of the kind that describe either the left- or right-going Henry are considered together, the ensemble has the form:

$$\rho = \frac{1}{2}\sum_j \left(| \leftarrow><\leftarrow |+| \rightarrow><\rightarrow | + e^{i\delta_j t}| \rightarrow><\leftarrow | + e^{-i\delta_j t}| \leftarrow><\rightarrow |\right)$$

The first two terms are independent of j and describe the probability of being in state $| \leftarrow>$ and $| \rightarrow>$, respectively. The last two terms, depending on the random phases δ_j, can be denoted as:

$$\rho_{\rightarrow\leftarrow}(t) = \rho^*_{\rightarrow\leftarrow}(t) = \frac{1}{2}\sum_j e^{i\delta_j t}$$

These terms are known as the *coherences* of the density matrix. In Figure 12.4 it can be seen that while each (jth) contribution to the coherences oscillates in time with its full amplitude, the sum over all contributions decays to zero due to their random phases (Figure 12.3).

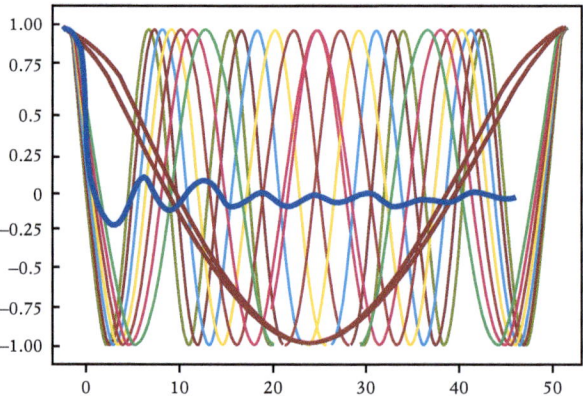

Fig. 12.4 Combined amplitude of many randomly detuned oscillators as a function of time (thick blue curve).

The combined amplitude undergoes damped oscillations until it settles at zero (the ground state). This behavior can be described by a *differential equation* for the coherences of the density matrix:

$$\frac{d\rho_{\rightarrow\leftarrow}(t)}{dt} = -R_{\rightarrow\leftarrow}(t)\,\rho_{\rightarrow\leftarrow}(t)$$

This expression is reminiscent of the *decay* rate of the probability amplitude due to its interaction with the environment (Chapters 9–11). This similarity shows the rapport between the two types of decoherence, known as *decay and dephasing*. The former pertains to *transitions* between the excited and the ground energy states that result in the decay of the *excited*-state probability amplitude, while the latter is associated with changes in the phase of *a state*, resulting in the decay of the *coherence terms between states*.

In the case of dephasing, after enough time has passed such that the phases are completely scrambled, the density matrix of the *ensemble* is given by:

$$\rho = \frac{1}{2}\left(\,|\leftarrow><\leftarrow| + |\rightarrow><\rightarrow|\,\right)$$

This is the completely *incoherent* state of a system with two orthogonal quantum states. Hence, random dephasing results in an ensemble without quantum coherence, that is effectively classical.

This phenomenon that arises in large ensembles of quantum systems is combatted by bang-bang or spin echo control. Repeated, periodic, flipping of phases in the entire ensemble eliminates the random phases at particular time instants, regardless of the phase values, thus rendering the *entire ensemble* coherent again.

PART III
Quantum complex systems and technologies

Henry Goes Through Walls

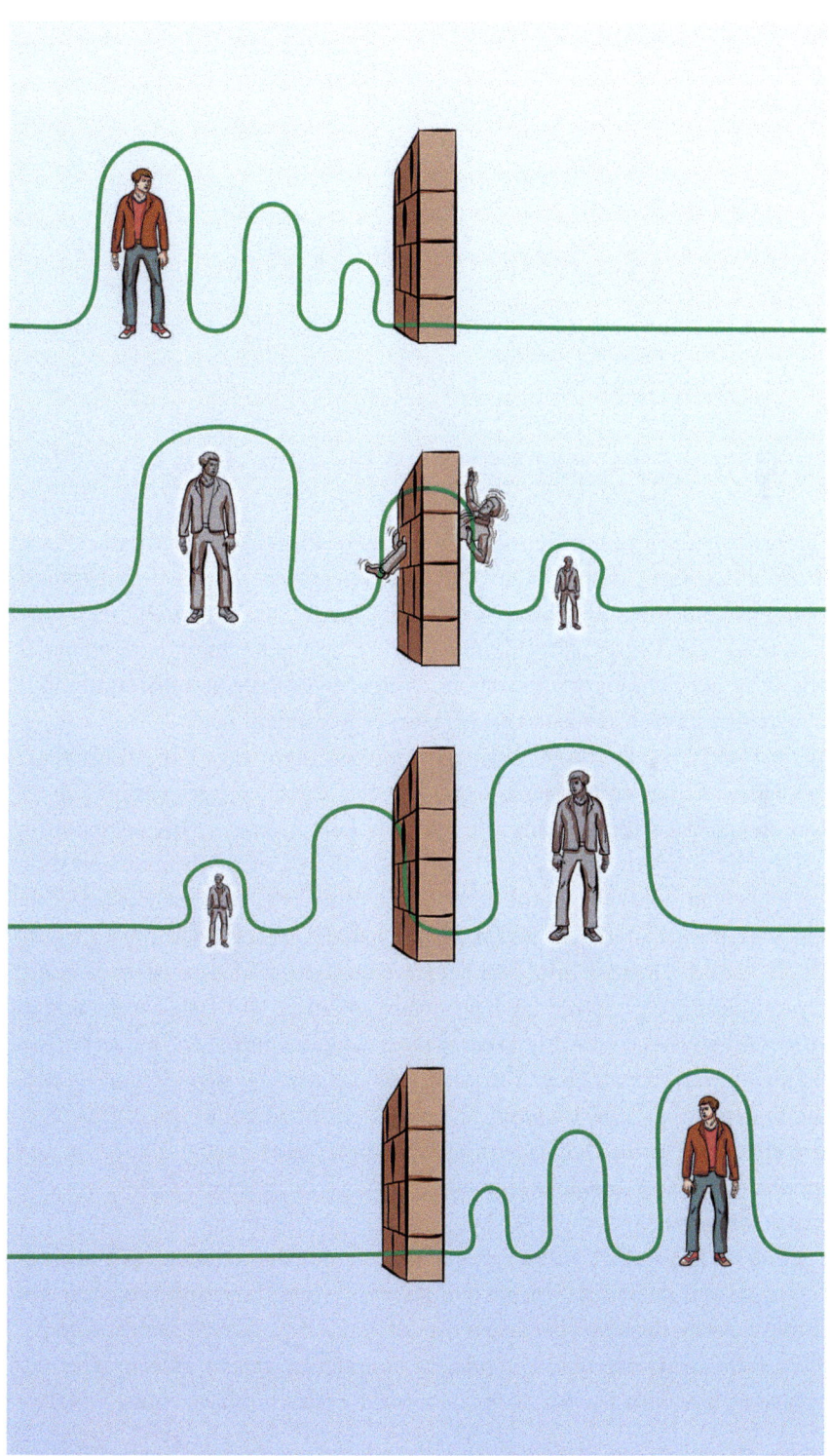

CHAPTER 13

What is Quantum Tunneling?

13.1 HENRY CHALLENGES IMPENETRABILITY

The gloom in the cell where Henry and Eve have been locked behind bars reflects their mood, close to despair. Can they expect clemency from their captors? Suddenly, an old memory stirs Eve's mind: "Henry, do you remember how we first met?" she whispers, trying not to wake up the dozing guard. Henry is surprised by her seemingly untimely reminiscence of their first encounter, but soon he realizes its relevance to their present predicament.

Henry vividly recalls the incident that brought them together twelve years earlier. He was then employed as a student at a classified laser fusion lab. As he was struggling with focusing the multiple laser beams of his machine on the target, he accidentally stepped into the focal area of the beams, and then . . . he found himself lying on the floor of the office next door, where a female student was sitting at the desk. "Are you in the habit of entering without knocking first?" she asked. "Forgive me, but if I haven't gone mad, I must have come in through the wall", he mumbled, completely baffled. "It's called tunneling in the quantum physics course I'm taking" she said, quite amused. "But aren't you rather big for that sort of thing? You don't look like an alpha particle that tunnels out of an atomic nucleus, do you?" she chuckled, leaving him speechless. This bizarre encounter sparked off their friendship and eventual collaboration, aimed at understanding and possibly recreating Henry's extraordinary experience of quantum tunneling.

A prime example of quantum tunneling is the radioactive decay of a nucleus that ejects an alpha particle, which is composed of two protons and two neutrons. Prior to the decay, the alpha particle is part of the nucleus, which itself consists of bound protons and neutrons. The nucleus boundaries act as a reflecting barrier that prevents the alpha particle from escaping if it does not have enough energy.

Were the particle classical, it would remain forever trapped in the nucleus. Why should a quantum particle act differently?

Recall that a quantum object is never fully localized in space: Due to the uncertainty principle (Chapter 5), its position is the more "smeared" out in space the less uncertain its momentum. Equivalently, the position smearing is determined by the spread of its quantum wavefunction. In several of his adventures, Henry's quantum suit allowed him to be delocalized or superposed over several locations (Chapter 2). There is, however, a new twist in the present adventure we need to address. What happens to a quantum wavefunction when it encounters a wall; i.e., a classically impenetrable barrier? Does it vanish at the wall boundary? A careful analysis of this problem by George Gamow (Sections 13.2 and 13.4) led him to the conclusion that although such a wavefunction would be predominantly back-reflected by the wall, it would still have a tiny "tail" that would penetrate inside the wall, albeit with an amplitude that exponentially diminishes with the wall's thickness (Figure 13.1). This small penetrability into a classically forbidden region was termed *quantum tunneling* by Gamow. Remarkably, despite its smallness, this penetration amplitude is clearly observed: its square is the tunneling (escape) probability of the alpha particle out of a single decaying nucleus, and we can infer from this probability the radioactive decay rate of the element in question. Related tunneling effects are common for nanoscopic, atomic or subatomic objects (Section 13.4).

But can one conceive of a similar effect for a macroscopic object such as Henry? Such a possibility appears to be highly unlikely, since the tunneling probability of a quantum object diminishes as its mass and the barrier thickness grow. One can estimate that the probability for Henry to tunnel through a few-inch thick wall is so small that the age of the universe would not suffice to see it happen with high probability. Nevertheless, highly unlikely events may still occur: they are known

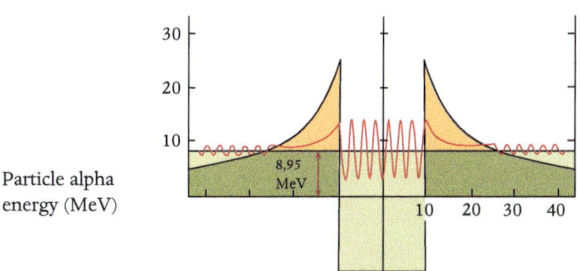

Particle alpha energy (MeV)

Fig. 13.1 The scheme of alpha decay proposed by G. Gamow in 1928. The wavefunction of the alpha particle (red) is confined by the nuclear potential, but tunnels through the potential barrier as an evanescent wave.

as *miracles*. Was Henry's experience a miracle? Perhaps, but several mitigating factors could greatly assist Henry's tunneling and make it much more likely.

Henry, as opposed to an alpha particle, did not escape into open space, but rather to another closed space, Eve's office. As a result, Henry's wavefunction bounces back and forth between the two rooms. The wavefunction acquires the shape of a *standing wave*, similar to a guitar string, clamped at its midpoint, that oscillates at a resonant frequency. A double-room space is similar to a double-well potential where the two wells are divided by a barrier (Figure 13.2(b)). The tunneling determines the tiny amplitude that couples the wavefunction portions which are localized in the two wells. Yet it is possible to enhance this coupling amplitude. Indeed, the multibeam high-powered laser that accidentally hit young Henry drastically increased the coupling between his two localized wavefunction portions, enabling him to tunnel into Eve's office.

The way to strongly enhance the tunneling probability is to drive either the object or the barrier by a periodic force at the tunneling resonance frequency: the energy difference between adjacent states of the double-well potential (Figure 13.2(b)). Such a process would be akin to the coherent Rabi transfer between two energy states that Henry made use of in his journey to the mine and back (in Chapter 9).

Another way to increase the transfer probability would be to frequently measure the energy of the object while it is trying to tunnel. These measurements should be repeated at a rate that would give rise to the anti-Zeno effect

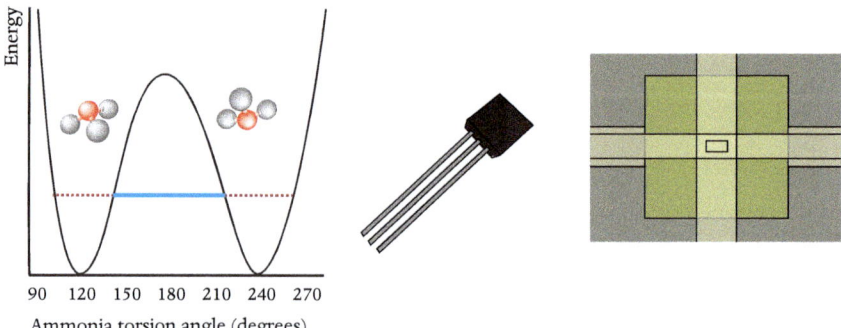

Fig. 13.2 (Left) Tunneling via a barrier between the two wells of a double-well potential is the operation principle of the first quantum radiation (microwave) amplifier, known as a *maser*, based on ammonia molecules. The double-well potential describes the energy of two possible molecular configurations that differ by their torsion angle. (Center) A semiconducting transistor, and (right) a superconducting Josephson junction, both of which rely on quantum tunneling for their operation.

(AZE) of enhanced transfer between the wells, similarly to AZE enhancement of decay probability (Chapter 11 and Section 13.2).

Both periodic driving and the AZE via periodic measurements can in principle render the tunneling rate large enough to be observable. A combined strategy can be adopted. First, the object is excited to an energy level as close as possible to the barrier top, so as to maximize the tunneling rate. Then, driving and/or measurements are applied at a rate that matches a tunneling resonance, thereby further enhancing the transfer between the wells. This strategy may be accomplished by careful tailoring of laser pulses. Young Henry was ignorant of the possibility of such a strategy at the time, yet by an incredible coincidence his fiddling with the powerful laser produced the right conditions for drastic enhancement of his tunneling into Eve's office.

Now, twelve years on, Henry has mastered the principles of quantum technology to the extent that Eve's allusion to his inadvertent tunneling incident makes him promptly realize what can get them out of captivity. Since the jail bars are a much less formidable barrier than a solid wall, the required tunneling enhancement may be realized by his quantum suit, which luckily is energized by their newly found quantum crystal that Henry managed to stash in his pocket. He then activates his rocket excitation mode combined with the Rabi transfer mode, similar to the ones he used for his journey to the mine, but now tuned to the tunneling resonance frequency of the jail bars. The quantum crystal radiates a series of intense pulses that make the jail bars oscillate at that frequency and simultaneously excite Henry's wavepacket to the maximal energy possible. Henry then tries time and time again to tunnel through the bars, until . . . hurrah! He is out! Now, while the guard is still snoring peacefully, there is a chance for Henry and Eve to sneak away to the rendezvous point from which Johnny's crew can fly them out upon request.

We skip the hair-raising details of their arduous and risky journey back home. Finally, their mission is accomplished. Together with Prof. Raman they shape the powerful quantum crystals they have so gallantly acquired into their long-sought device: the quantum super-lens. This lens should enable new, fantastic quantum technologies that, as our protagonists believe, will precipitate the quantum revolution. But nasty surprises lie ahead . . .

13.2 QUANTUM TUNNELING AND WAVEPACKET INTERFERENCE

Henry's and Eve's latest adventure sheds further light on a fundamental aspect of QM which we dealt with in Chapters 2 and 3: wavepacket interference. The new twist in this chapter is tunneling: the ability of quantum wavepackets to penetrate into classically forbidden regions. Tunneling was theoretically discovered in 1928 by G. Gamow (a Russian émigré who settled in the US) and independently by R. Gurney and E. Condon. This effect explained the breakup of radioactive atoms through the emission of alpha particles (comprised of two protons and two neutrons). The idea of tunneling is that a quantum particle (an alpha particle in the experiments analyzed by Gamow) is allowed to escape from the region where it is initially confined, though such escape is classically forbidden because of the lack of energy by the particle. In Figure 13.1, the particle escapes from a narrow region of low potential energy (a potential well) inside the atomic nucleus through a narrow region of high potential energy (a potential barrier). What makes this effect exclusively quantum mechanical is that according to Heisenberg's position–momentum uncertainty relation (Chapter 5), the confinement of the particle in a narrow potential well allows for the possibility of the particle having sufficiently large momentum, and consequently kinetic energy, for crossing the classically forbidden barrier (Figure 13.1). Yet this explanation is somewhat simplistic, as it does not reveal the dynamics of the tunneling process. We may devise an explanation in the spirit of Chapter 10, whereby the particle in the well occupies an unstable energy state which decays at a low rate through the barrier to an empty "bath" (equivalently, an unpopulated energy-state continuum). The decay rate through the barrier is very low because it is determined by the tiny spatial overlap between the confined wavefunction in the potential well and the extended free-particle wavefunction outside the barrier (Figure 13.1): the thicker and higher the barrier, the lower the decay rate (or, conversely, the longer the lifetime of the confined particle).

Following its success as an explanation of alpha-radioactivity, it has turned out that tunneling is a general phenomenon that applies to a broad variety of processes, including the beta decay of a neutron in a radioactive nucleus, electron tunneling in semiconductor junctions that underlies the transistor action, electron-pair tunneling that enables superconducting (Josephson) junctions, and electron tunneling between the two wells of the molecular potential in ammonia that is the operation principle of a quantum radiation (microwave) amplifier known as a maser, to name but a few (Figure 13.2). In recent years, tunneling of rather large and complex multi-atom systems has been observed, though they are not quite as large and complex as Henry in our story. The time may come, however, when our fantastic story will not be too far-fetched . . .

Let us start our explanation of the physics that underlies tunneling by recapitulating Gamow's analysis: He explained the fact that the tunneling rate depends on the spatial overlap of the initial (confined) and final (quasi-free) wavefunctions by matching the solutions of the Schrödinger equation inside and outside the barrier. This matching shows that a small "tail' of the confined wavefunction penetrates through the classically forbidden barrier—a tail that decreases exponentially as the barrier widens. The same exponentially decreasing tail of the wavefunction is obtained by adding an imaginary part to each component of the wavefunction endowed with a specific, real, momentum (Section 13.4). Such imaginary-momentum addition to the real momentum gives rise to mathematically valid solutions of the Schrödinger equation for tunneling setups, and is fully analogous to evanescent-wave solutions at interfaces between media with different opacity to light (Figure 13.3), which were discovered in the context of wave equations for classical light well before the advent of QM. Yet such imaginary or evanescent wave solutions are not physically transparent (pun intended) because physical observables such as momentum must have real eigenvalues (Chapter 4).

There is, however, an approach that reveals the physical nature of tunneling wavepackets. This approach, developed by R. Feynman (USA) in the 1940s, is called the path-integral formulation of QM. Its starting point is Schrödinger's observation that a quantum particle, in contrast to its classical counterpart, does not follow a well-defined trajectory or path in time, as such a path would contradict the Heisenberg position–momentum uncertainty relation. Instead, Feynman postulated that the evolution of a quantum particle is determined by the simultaneous propagation of the initial wavepacket along all possible paths leading to a common end point, with each path weighted by its appropriate

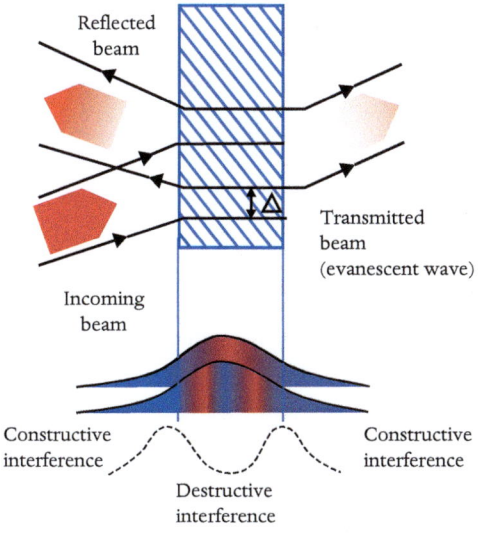

Fig. 13.3 Evanescent waves at interfaces between media with different opacity to light.

probability amplitude. Therefore, we must allow not only for those paths that the particle has enough energy to traverse according to classical mechanics, but also for all other paths that are classically forbidden. The sum over paths weighted by their probability amplitudes constitutes the same interference patterns discussed in Part I, such as the splitting and recombination of wavepackets in Henry's adventures (Chapters 2 and 3).

Feynman's approach may be invoked to interpret tunneling as predominantly destructive interference between the wavepacket portions that are peaked inside the barrier, resulting in strong overall suppression of the wavepacket in that region. Concurrently, the forward tails of the same wavepacket portions that extend beyond the barrier may have appropriate phases to undergo constructive interference. As a result, these forward tails may reinforce each other and penetrate much further through the barrier than a wavepacket following a single trajectory in this region (Figure 13.4).

Feynman's approach may also explain another anomaly: the penetration through the barrier appears to be much faster than light! This anomaly was

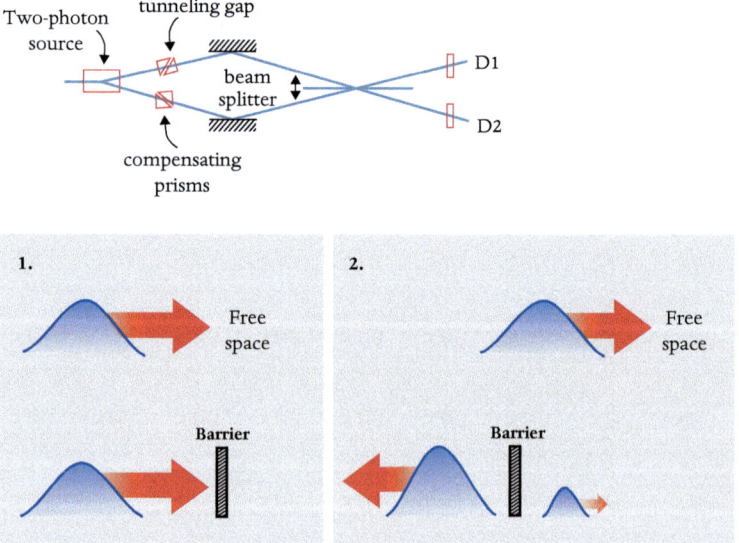

Fig. 13.4 The Chiao–Steinberg–Kwiat setup for measuring the path-length difference of photon pairs that are simultaneously produced in a crystal. One photon travels through free space, whereas the other, traversing the *same* distance, encounters a barrier realized by a grating, as in Figure 13.3. In the rare events in which the latter photon tunnels through the barrier, thus allowing two-photon coincidence to be recorded, the path traversed by the tunneling photon turns out to be *shorter* than that of its free-space counterpart. Can one then claim that the tunneling photon propagates faster than in free space; i.e., with superluminal speed? See text.

demonstrated by R. Y. Chiao, A. Steinberg and P. Kwiat (USA) in 1990: their light source produced pairs of photons, one of which traveled through empty space, while the other one traversed the same distance through a setup that acted as a tunneling barrier (Figure 13.5). In most experimental runs, the photon on the upper path on the figure failed to penetrate through the barrier and was back-reflected, but in the few runs where the photon tunneled through, it overtook by far the photon on the lower path of the figure that went through empty space. The detector that recorded the arrival of the tunneling photon had to be moved further away than its counterpart that responded to the arrival of the other photon in order to achieve *simultaneity* of the two photons. Does this anomaly violate the venerated causality principle that follows from Einstein's special relativity which forbids faster-than-light (*superluminal*) signaling? If this were the case it would spread havoc in the world: superluminal signals could be used, as in many time-machine science-fiction stories, to create logical paradoxes.

Luckily, there is a way around such paradoxes, as shown by Y. Japha and G. Kurizki (Israel, 1995–98) based on Feynman's approach. Each photon propagates in the barrier as the wavepacket shown in Figure 13.3, and thus its portion that tunnels through the barrier is reshaped: the forward tails of the split wavepacket portions in the barrier region interfere constructively, yielding an apparent peak of the entire wavepacket as it exits the barrier. Since this apparent peak arrives at the detector much earlier than the wavepacket peak in free space, it is tempting to ascribe faster-than-light (superluminal) speed to the tunneling photon (Figure 13.4). However, this early-arriving peak has a *smaller probability* of being detected than the forward tail of the wavepacket in free space, because it is

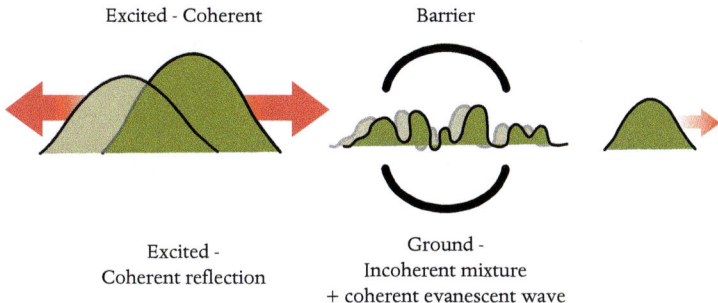

Excited - Coherent   Barrier

Excited -   Ground -
Coherent reflection   Incoherent mixture
                             + coherent evanescent wave

Fig. 13.5 Tunneling control in a barrier where an excited-atom wavepacket is subject to photon emission, which blurs the atom energy and destroys its coherent interference. The resulting ground-state wavepacket is incoherent and acts as a mixture, and its less energetic portion is back-reflected while the more energetic portion goes through the barrier as a classical particle would. Dynamical control of the photon emission can either suppress this decay and the resulting incoherent wavepacket spread (in the QZE regime), or enhance it (in the AZE regime).

attenuated by the multiple reflections at the barrier. Hence, the apparently superluminal photon tail does not signal before its empty-space counterpart. Equivalently, we may consider an ensemble of many photons launched towards the barrier. Then, only an exponentially small fraction will tunnel through, producing fewer detector clicks at early (superluminal) times than their counterparts in a free-space ensemble of the same number of photons, if we allow for the spread of arrival times dictated by the time–energy uncertainty of the free-space wavepackets (Chapter 6).

The interpretation of tunneling as coherent wavepacket interference inside the barrier implies that decoherence would destroy tunneling; that is, it would turn the wavepacket that tries to penetrate the barrier into an incoherent mixture of wavepacket portions that behave nearly classically. These mixtures cross the barrier if their mean energy exceeds the barrier potential energy or are back-reflected otherwise (Figure 13.5). This property may be used to dynamically control tunneling through a barrier region that is coupled to an environment by methods akin to the QZE, the AZE or dynamical echo (bang-bang, BB) control discussed in Chapters 11 and 12. Since the QZE and BB control are aimed at suppressing the environmental decay and decoherence, they can, if successful, restore coherent interference at the barrier, thus giving rise to tunneling in its environment-free form. By contrast, AZE control that enhances decay and decoherence may totally destroy tunneling by rendering environmental effects dominant and obliterating coherent interference in the particle propagation through the barrier.

An alternative view of these decoherence effects is that dynamical control incurs energy spread into the particle's motion. This spread blurs the distinction between classically forbidden tunneling and classically allowed propagation by smearing the mean energy of the wavepacket such that it is uncertain whether it is below or above the barrier. Dynamical control can either drastically increase or decrease the probability of barrier penetration, depending on the interplay between the energy spread of the particle caused by the environment, and by control.

13.3 MOTION AND ITS LIMITATIONS IN QUANTUM MECHANICS

In a popular essay, Schrödinger identified the essence of QM with the breakdown of the trajectory concept. He thereby focused our attention on the *unpredictability of the future location* of a quantum particle given its present (observed or measured) location. This unpredictability can stem from the position–momentum uncertainty relation that pertains to single-particle quantum wavepackets. If the wavepacket propagates through a medium that causes multiple nearly-complete reflections as in the case of tunneling, then the coordinate spread grows with each reflection, making its future location even more uncertain. Feynman's approach, whereby a quantum particle propagates via multiple interfering trajectories, is an equivalent statement concerning the breakdown of the trajectory concept in QM.

A variety of effects reveal the "paradoxical" nature of motion in QM, in the sense that they have no classical counterparts. Thus, in the case of two entangled particles, the uncertainty of each particle depends on the precision of measuring its counterpart. In particular, the method termed *ghost imaging* (Figure 13.6), proposed by D. Klyshko (Russia, 1995) and experimentally verified by Y. Shih (USA, 1995), demonstrates the ability to manipulate the wavepacket of an "absent" (undetected) particle by taking advantage of the spatial correlation between entangled partners. The QZE and AZE (Chapter 10) may drastically change the localization of a quantum particle; i.e., exclude it from a certain region or confine it there, as well as alter the arrow of time of a particle by interchanging its past and future. Finally, extremely frequent measurements at intervals of 10^{-20} sec or less would incur a huge energy spread, larger than 1 million electron

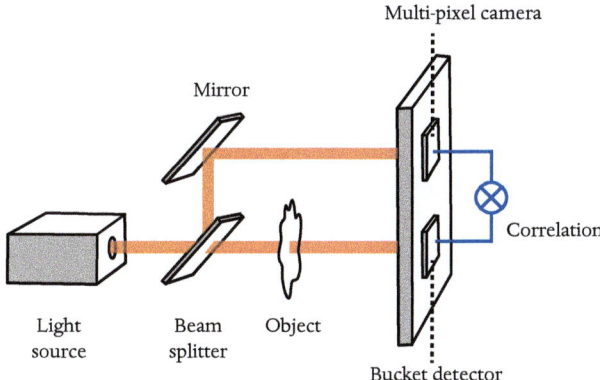

Fig. 13.6 Ghost imaging. An image is recorded by the upper (multi-pixel) camera only if the lower "bucket detector" clicks. Thus, a *silhouette* of the object is recorded by the upper camera, even though the photons on the upper path have never interacted with the object.

Volts (1 MeV) that may create new particles—i.e., electron–positron pairs—out of the vacuum (Chapter 11).

The exotic effects discussed here imply that QM might have fundamentally altered not only the concept of motion but also our entire grasp of space, time and being, which Kant (Prussia, eighteenth century) viewed as immutable categories of our perception of the world. Quantum aspects of cosmology suggest an answer to the issue discussed since the dawn of Greek philosophy—an issue that underlies Zeno's paradoxes (Chapter 10). Are space and time divisible into ever-decreasing intervals without any bound (*ad infinitum* in Latin)?

The answer may be given by the notion of Planck's length, 10^{-43} cm—the smallest spatial interval allowed by QM and general relativity—if the two are to agree. Intriguingly, the Planck length is also the smallest size to which a black hole can shrink by evaporation via the emission of Hawking radiation. This length scale is also surmised to be the size of the primordial universe at the instant of the Big Bang: thus, the initial size of the universe is the ultimate size of its smallest objects, below which the notion of space is meaningless.

The shattering of the deeply rooted categories concerning space, time and causality may profoundly affect human existence. It is possible that new categories or paradigms that will replace them may reshape our fundamental notion of what is small and what is large, and help us bridge distances in space and time in ways yet unimaginable. The anticipated new paradigms may also confront us with observer-dependent structure of space and time and our ability to control these key aspects of our existence. Will such omnipotent ability make us even more self-conscious and arrogant than we are now? Perhaps such danger may explain why Jewish sages who authored the Mishna (Palestine, second century AD) forbade the posing of the following questions. What is above (beyond the world), and what is below (underneath the world)? What had been before the world came to be, and what will be after it ends?

The Wall

How often do we face a sturdy wall
That bars our life route, our aspirations
And tempts us to forsake them all,
Succumb to listlessness and desperation?
Now science inspires us: not all is lost!
A mighty barrier may be challenged still
By finding its weak point and, at high cost,
Assaulting till it yields to our will.

13.4 APPENDIX: TUNNELING AND THE SCHRÖDINGER EQUATION

Until now, when considering Schrödinger's equation (as in Sections 6.4 and 9.4) we have only discussed the time-dependence of the wavefunction, but not its variation in space under given constraints in energy and potential shape. Here we discuss such spatial variation in the case of quantum tunneling.

Quantum tunneling occurs when there is a spatial change in potential energy, from a low potential energy (the prison cell or Henry's past lab) to a much higher potential energy (the prison bars or the wall). Henry's energy determines where he can be *classically*: if his energy is higher than the potential energy, then he can roam about freely, but he cannot (classically) penetrate a "potential barrier"—a region where the potential energy is higher than his own. Therefore, if there is a wall that requires more energy to climb than Henry has, then, classically, he cannot succeed.

However, as we have learnt from Henry's last adventure, quantum mechanics has a different notion of what is possible and what is not. In order to address this scenario, we need to introduce Schrödinger's equation in its spatially one-dimensional time-independent form, known as the *wave equation*:

$$-\frac{\hbar^2}{2m}\frac{d^2\psi(x)}{dx^2} = [E - V(x)]\,\psi(x)$$

This equation may seem daunting, but we have encountered each of its constituents before: $\hbar$ (Henry's insignia) represents Planck's constant; m is the mass of the object (in our case Henry's); and $\frac{d^2\psi(x)}{dx^2}$ is the *second derivative* of the wavefunction with respect to the spatial coordinate x, which shows how rapidly the wavefunction's first derivative changes along x, for example, from the prison cell through the bars to the outside. On the right hand-side of the equation we again have $\psi(x)$, the wavefunction. Thus, this equation represents a relation between the way a wavefunction changes in space and its value at each point of space x. This relation is dictated by the term $[E - V(x)]$, where E is Henry's energy, and $V(x)$ is the variation of the potential energy around Henry. Wherever $E > V(x)$, Henry has more energy than the potential energy, so that he can move around freely. The question we now pose is: what happens when $E < V(x)$; i.e., how does Henry's wavefunction look like in places where he does not have enough energy to *be found classically*?

The solution to this equation, in its most general form, is extremely complicated, so we will consider a simplified version in which the potential energy,

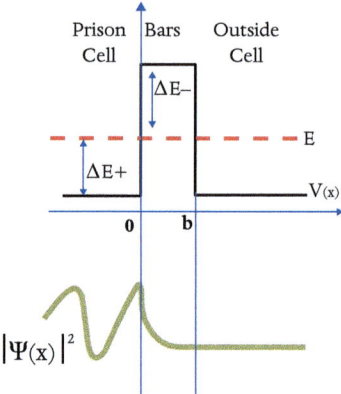

Fig. 13.7 A depiction of the energy landscape (upper panel) and the wavefunction (lower panel) as a function of x.

$V(x)$, is piecewise-constant; that is, it has a constant value within the cell, a higher constant value within the bars, and again a smaller constant value outside the cell (Figure 13.7). For a constant potential, V, the solution to the equation is obtained from the relation $\frac{d}{dx} e^{kx} = k e^{kx}$ and becomes very simple:

$$\psi(x) = A e^{ikx} + B e^{-ikx}, \ k = \sqrt{\frac{2m\,(E - V)}{\hbar^2}}$$

Here, A and B are constants that we will explain later, and k is the wavenumber which dictates the spatial scale on which the wavefunction changes. There are two qualitatively different regimes. First, $E > V$. Let us define $\Delta E_+ = E - V$, which is proportional to the expression inside the square root. When this expression is positive, the wavefunction behaves as a sum of running waves e^{ikx} propagating to the right, and e^{-ikx} propagaying to the left, where $k = \sqrt{\frac{2m\Delta E_+}{\hbar^2}}$. If $A = B$ these running waves interfere to form standing waves, oscillating in space as $\cos kx$, and if $A = -B$, as $\sin kx$, as we have seen before (Section 3.4). These standing waves describe a quantum particle reflected by the outer boundaries of the potential. Thus, when Henry has more energy than the potential, his wavefunction behaves as that of a free particle: He can roam freely within the bounds of space where $E > V$. Second, $E < V$. Let us define $\Delta E_- = V - E$. In this scenario, $k = iq, q = \sqrt{\frac{2m\Delta E_-}{\hbar^2}}$, where $i = \sqrt{-1}$ came out of the square root. Now, the wavefunction amplitude behaves as a sum of e^{qx} and e^{-qx}. Due to the normalization constraints of the wavefunction, it cannot explode to infinity and hence e^{qx} has no physical meaning. The only physically admissible wavefunction amplitude diminishes within the restricted area with negative energy difference, as e^{-qx}. While this amplitude may be extremely small if $qx \gg 1$, it is not zero for finite x. Hence, Henry's wavefunction has a non-zero probability amplitude within the bars.

We can now write the entire solution for Henry's wavefunction to an adequate approximation, as follows:

Within the prison cell,

$$x \leq 0 : \psi(x) = e^{ikx} + B_{cell}e^{-ikx} \quad \text{a wave traveling right and left}$$

Within the bars,

$$0 \leq x \leq b : \psi(x) = B_{bars}e^{-qx} \quad \text{a decaying amplitude}$$

Outside the prison cell,

$$x \geq b : \psi(x) = A_{free}e^{-ikx} \quad \text{a wave traveling to the right}$$

In order to determine all the As and Bs we employ the physical requirement that the wavefunction $\psi(x)$ and its spatial derivative be continuous. This yields constraints at the border between the different areas:

$$\text{I. } x = 0, \psi(x) : 1 + B_{cell} = B_{bars}$$

$$\text{II. } x = 0, \frac{d\psi(x)}{dx} : ik - ikB_{cell} = -qB_{bars}$$

$$\text{III. } x = b, \psi(x) : B_{bars}e^{-qb} = A_{free}e^{ikb}$$

$$\text{IV. } x = b, \frac{d\psi(x)}{dx} : -qB_{bars}e^{-qb} = ikA_{free}e^{ikb}$$

The solution to these four equations with four variables is relatively simple. The important question is: what happens to the free part; i.e., can Henry tunnel through the bars and outside the cell? The probability amplitude of the wave-function outside the cell is proportional to A_{free}, which from III is proportional to e^{-qb}:

$$|\psi(x)|^2 \sim e^{-2qb}$$

Thus, the thicker the bars (the larger b) and the harder they are to pass (the larger ΔE_-), the lower is Henry's probability of breaking free from the cell. In our scenario, this probability is negligibly small on account of m, Henry's mass, being large. However, according to our explanation in Section 13.1, Henry's quantum suit energized by a quantum crystal has a hugely enhanced ability to penetrate this barrier, owing to the large, controlled amount of energy the crystal provides, thereby drastically reducing ΔE_-.

To conclude, a potential barrier, which classically would be impenetrable, still allows in quantum mechanics the possibility of tunneling; i.e., there is a probability amplitude to penetrate the barrier, albeit small. As discussed in the previous sections, tunneling is actually ubiquitous in phenomena such as radioactive decay, and is the key to devices such as semiconductor transistors, superconductiong quantum interference devices or tunneling electron micro-scopes, to name just a few (Figure 13.2). Henry's ability to implement this effect on a macroscopic scale is, however, still beyond our reach.

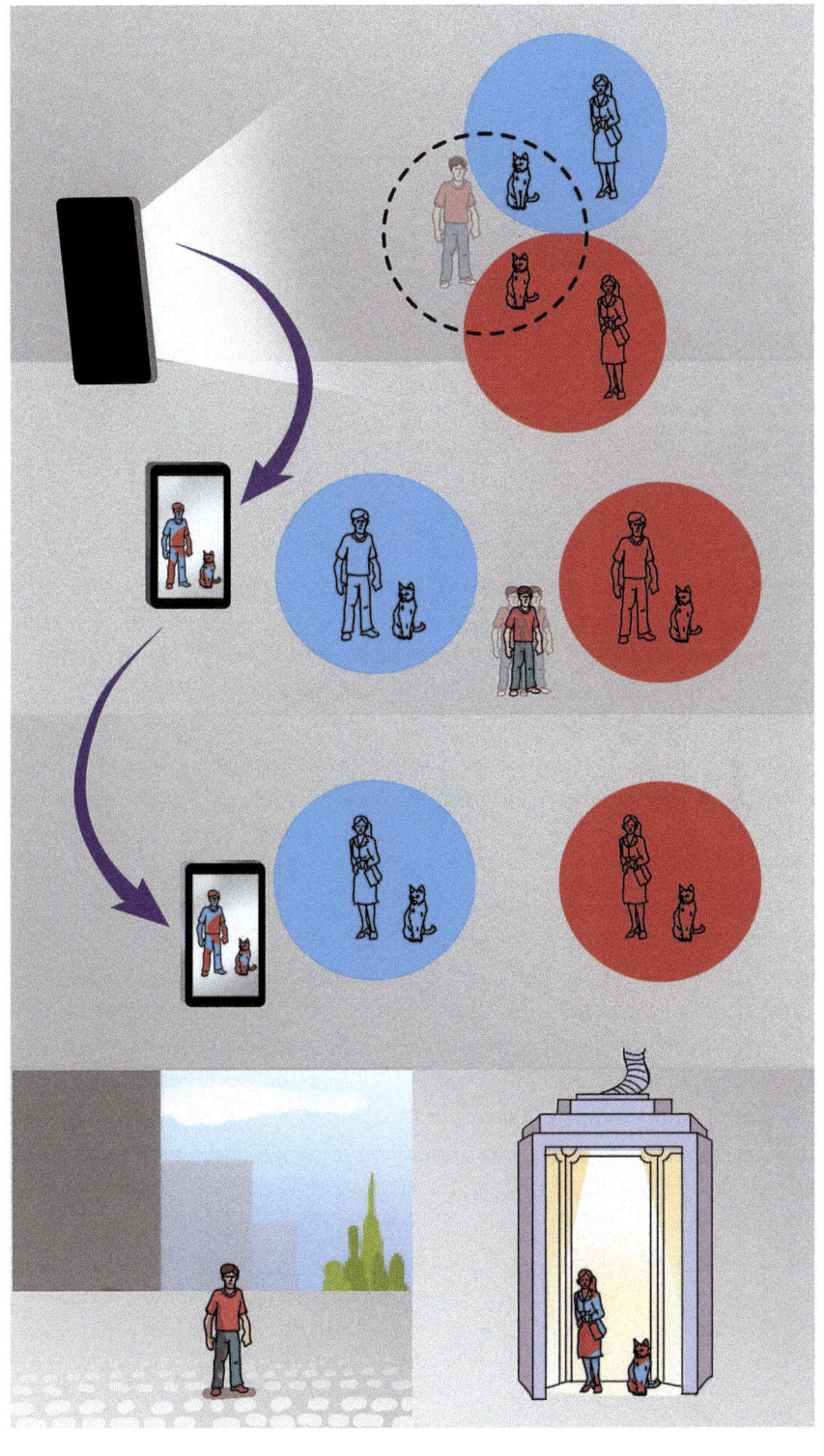

CHAPTER 14

What is Quantum Teleportation?

14.1 TELEPORTATION TRIO

The powerful quantum super-lens that Henry, Eve and Prof. Raman have crafted has allowed them to create formidable new quantum devices, thanks to the variety of functions this super-lens combines: generation of quantum correlations, alias entanglement, between quantum systems (Chapter 7), their protection from decoherence or disentanglement by coherent control (Chapter 12) and their quantum measurements (Chapter 4).

Confident that their new line of quantum devices is close to completion, Henry, Eve and their loving cat Schred celebrate their success at the lab, all three of them wearing their quantum suits with all their new quantum gadgets installed. To commemorate the occasion, Eve entangles herself to Schred before taking off to the Quantum Conference at Waikiki, Hawaii, where she is due to present their latest achievements.

The next day, a bizarre and ominous incident befalls Eve as she, in a split quantum state, is about to enter the Conference venue: a menacing thug on a motorcycle tries to run her over, or perhaps kidnap her. Eve just barely escapes this assault and, still fearing another attempt on her life, calls Henry. Alas, Henry is too far away to be of help—or is he? He realizes that the only chance is to try out their newest invention, the quantum teleportation chamber, which he enters swiftly together with the split Schred . . . And then, a wonder! A split-second later, the thug is facing Henry armed with a bat instead of Eve! The standoff ends as the thug backs off and dashes away on his motorcycle. Meanwhile, in Eve's and Henry's hometown, thousands of miles from Waikiki, Eve and Schred emerge from the teleportation chamber. Our protagonists can rejoice at their success: the first demonstration of quantum teleportation on live creatures, whereby Henry and Eve, aided by Schred, have traded places *in virtually no time at all*, despite their huge distance (thousands of miles) apart!

Let us explain the essence of the quantum teleportation protocol. Its goal is to teleport—to transfer—a quantum *state* from one system to another, regardless of how distant they are. The crux of the teleportation protocol is that initially Henry was in his own state and Schred and Eve were in an entangled state, while at the end of the process, Henry and Schred are entangled and Eve's matter receives Henry's state (thus *becoming Henry*), whereas Henry's matter receives Eve's entangled state with Schred (thus *becoming Eve*).

The protocol requires a quantum-entangled channel composed of two systems in a known entangled state. One of them is an auxiliary system located near the sender of the quantum state, and the other, located at the receiving end, is the quantum state receiver. In our case, Schred and Eve form the quantum-entangled channel, wherein Schred is an auxiliary system colocated with the sender, Henry, while Eve is at the receiving location. Henry and Schred enter the teleportation chamber, where they are subject to a joint measurement in the entangled-state basis of the human and the cat. Henry sends the measurement result to Eve's phone, and she rotates her state by an appropriate operator. As a result, Henry's state, which encodes all of Henry, appears where Eve's state was, and Eve appears at Schred's location, where they are entangled, as they were at the outset.

The chain of events in the teleportation protocol is as follows.

A) The protocol starts with a bipartite measurement of the quantum-state sender and the auxiliary system. The two are jointly measured in the basis of their entangled states. The measurement results in a collapse (projection) of their joint wavefunction, to an entangled state of the two systems. In our case, Henry and Schred are jointly measured in the teleportation chamber. Thus a measurement of *both Henry and Schred* is performed at the same time, unlike the measurement of one system at a time that we have encountered so far. Usually one measures in a separable basis, which would here allow us to ask whether Schred is here or there *or* whether Henry is here or there. By contrast, in the present joint measurement, which is performed in the entangled-states basis, the possible results are: 1) first entangled state (Schred is here, Henry is here) *and* (Schred is there, Henry is there); 2) second entangled state (Schred is here, Henry is there) *and* (Schred is there, Henry is here). There are two more entangled states whose details we skip (see Section 14.4). The result of the measurement is random, since it is unknown which of the four entangled states has been projected out, but the joint Henry–Schred state inevitably *collapses* to an entangled state following the measurement. This is utterly counterintuitive. Initially, Henry and Schred are *not* entangled; and then the measurement causes Henry and Schred to be entangled! By this measurement, entanglement has been *instantaneously* transferred from the quantum-entangled channel (here Schred + Eve) to the channel of the sender + auxiliary (here Schred + Henry).

B) After the measurement, due to the randomness of its results, the Henry–Schred entangled state is unknown *a priori*. Therefore, Henry takes a quantum-selfie that measures his entangled state with Schred. The measurement result is *classical* information transmitted via the cellular phone to Eve's quantum suit. This information enables a simple *local* quantum operation by Eve, turning her into Henry. Each of the four possible measurement results requires a different local quantum operation by Eve. Thus, remarkably, QM ensures that the entanglement transfer must entail the transfer of the sender's state to the receiver (here Henry → Eve), provided the measurement result is communicated from the sender to the receiver and the receiver performs a suitable operation determined by the measured result.

If the protocol has been perfectly performed, the receiver's state is then identical to the sender's state. Since only the *quantum state* dictates the system's properties, the teleportation literally *changes the receiver into the sender*. In our case, Henry's quantum-selfie, received by Eve's quantum suit, allows an operation that alters the quantum state of the receiver (Eve's atoms and molecules), replacing it by the *quantum state of the sender* (here, Henry's atoms/molecules). What is teleported is not *matter*, which is a substrate present at the receiving end of the teleportation link, but rather *information* embodied by the sender's quantum state.

14.2 QUANTUM TELEPORTATION AND CRYPTOGRAPHY

Two landmark experiments—the first by J. Clauser (USA, 1972) and the second by A. Aspect (France, 1982)—demonstrated that entangled photon pairs exhibit peculiar non-local correlations; namely, correlations that persist regardless of the distance between their source and their receivers. Such correlations do not exist for classical variables but only for joint, non-commuting variables of quantum-entangled systems. For photon pairs these are correlations between results of measurements performed by two distant observers on each of the orthogonal (x, y or z) components of the joint photon-pair polarization that are analogous to those of the joint spin of two electrons (Figure 14.1).

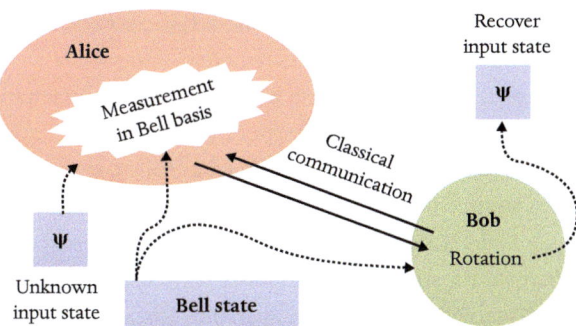

Fig. 14.1 Quantum teleportation protocol. Alice has a qubit in an unknown state. Alice and Bob share a pair of entangled qubits in a specific entangled Bell state. Alice performs a two-qubit Bell measurement and transmits to Bob her (classical) result. Bob then rotates his qubit according to that result and retrieves Alice's initial input state.

Clauser's and Aspect's experiments put an end, to a large extent, to the debate concerning whether QM allows for non-local correlations of joint two-particle observables, as opposed to the classical description of such observables. The corresponding distinction between results of quantum measurements performed on the two distant and entangled particles or photons, for which QM predicts correlations, and their classical counterparts, for which the results are expected to be uncorrelated, was quantified by J. Bell (UK, 1964), who actually did not believe that such a distinction exists in reality, just like Einstein (Chapter 7). The aforementioned experiments proved these "quantum skeptics" wrong. Curiously, Clauser was motivated by the opposite conviction: he shared the Hindu or Buddhist beliefs of some of the Hippie Generation that the world is one indivisible entity, and set out to seek the manifestation of this indivisibility in QM.

The resolution of this debate would have probably remained within the academic sphere had it not been for the appearance of two seminal theoretical papers. The first, by C. Bennet (USA) and G. Brassard (Canada) in 1984 (BB84 protocol) and the second by A. Ekert (UK, 1991), raised the possibility of employing measurements of entangled photon pairs by distant observers for secret-key distribution between the observers, which has since become the foundation of quantum cryptography, as described subsequently.

The other paper, coauthored by C. Bennett, R. Jozsa and W.K. Wooters (USA), G. Brassard and C. Crepeau (Canada) and A. Peres (Israel) in 1993, proposed the use of two-photon entanglement for quantum-state teleportation between distant locations—a notion that at first sight appears to resonate with science-fiction ideas of how to traverse cosmic distances in no time, but in reality implies nothing of this sort, as will be explained.

The common impact of these papers has been revolutionary in hindsight, as they have identified an entirely new perspective: unique technological applications enabled by quantum entanglement. In what follows we elaborate on each of the ideas expressed in these papers and their development over the years:

Quantum cryptography. Ekert's protocol for quantum cryptography requires a source of polarization-entangled photon pairs (operated by Charlie) that sends one photon in each pair through a polarizer to one observer (Alice). and the other photon to another observer (Bob), who detects them, also through polarizers (Figure 14.2). The encryption consists in the direction (orientation) at which the polarizers are set by the sender and its readout by each observer through her/his polarizer.

As in Aspect's experiment, both the sender and the receivers of each pair can set their polarizers at one of three orientations differing by 45° from each other, in order to use the polarization components as elements of the encryption key. Charlie encodes his secret key in a sequence of polarizer orientations. Having shared many pairs of entangled photons with Alice and Bob, who set their polarizers randomly in one of the three allowed orientations, it is their turn to compare their results via an open (public) channel. If their polarization is

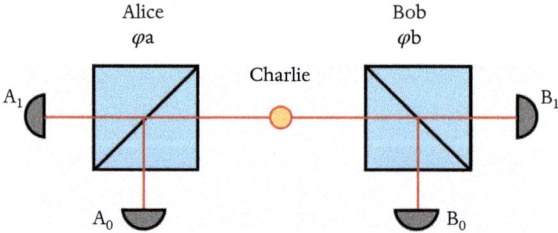

Fig. 14.2 A protocol for secret-key distribution, in which a sender (center circle) shares a polarization-entangled photon pair between two receivers, by sending one photon to Alice and the other to Bob. Each of them selects a polarization-measurement basis by rotating the polarizing beam-splitter, thereby adjusting the probabilities of measuring their respective photons at their detectors (A1 and A2 for Alice, B1 and B2 for Bob). Eventually they compare their results for their respective photon polarizations, and discard those events where their polarizer settings (rotation angles) did not coincide, as in the diagram, where the rotation angles of Alice and Bob differ. For the measurements with complementary rotation angles, the key is composed of 0s (a click in A_0 for Alice and B_0 for Bob) and 1s (a click in A_1 for Alice and B_1 for Bob).

parallel, they deem the result correct and use the measurement result (0 or 1) as the key. Otherwise they treat the result as an error and discard it (Figure 14.2). In the ideal case of fully entangled photons and error-free photon detection by Alice and Bob, the fraction of correct results measured along the three possible orientations reveals the elements of the secret key and confirms the photon-pair entanglement. Yet if Eve intercepts part of the photons trying to eavesdrop on the encryption and subsequently releases these photons to be detected by Alice and Bob, those photons will be disentangled. Hence the fraction of erroneous results by the two observers will be larger than expected, alerting them to the presence of an eavesdropper. This makes quantum cryptography a powerful, eavesdrop-proof method, but only in an ideal case and not in practice.

In reality, since the photon pairs are partly disentangled by decoherence or the detector efficiency is less than 100%, as is always the case, the resulting errors will be indistinguishable from those of eavesdropping. Later, additional quantum cryptography protocols were proposed and demonstrated experimentally. All these protocols are limited by decoherence and detector efficiency, but have nevertheless achieved great success. Encryption key distribution over 420 km in optical fiber or over 300 km in open air (between two of the Canary Islands) by an Austrian team, as well as from earth to a satellite in space (over 1,400 km) by a Chinese team, are the most spectacular achievements to date (Figure 14.3).

Quantum teleportation. The basic protocol of quantum teleportation by Bennett and coworkers is as follows (Section 14.4). Alice and Bob share a pair of entangled photons. Alice measures her photon jointly with an unknown photon (a polarization qubit, just like her other photon) and communicates the result (by a classical link, such as a phone) to Bob. Based on her result, Bob performs a unitary operation that rotates his photon onto the state of the unknown photon and

Canary   **Earth & Satellite**

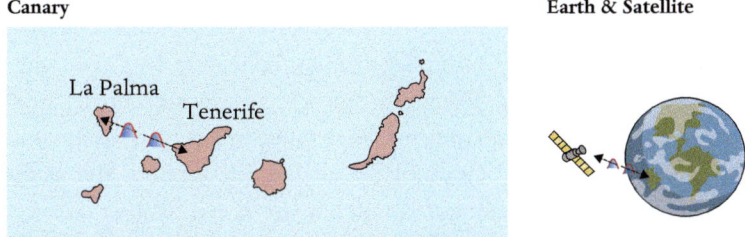

Fig. 14.3 Quantum teleportation and cryptography links by entangled photon pairs (colored wavepackets) have been experimentally demonstrated by an Austrian team between the Canary Islands of La Palma and Tenerife (left), and by a Chinese team between an earth-based station and a satellite in space (right).

thereby recreates the unknown state at Bob's location. Schematically, this means that the unknown state of particle 1 is teleported to particle 3 upon (a) sharing the entangled particles 2 and 3 between Alice and Bob, (b) jointly measuring the 1+2 particles by Alice, and (c) using the result of the measurement to impose the unknown state of particle 1 on particle 3.

Since the communication of the measured result occurs at the speed of light, there is no faster-than-light signaling involved in teleportation, much to the dismay of science fiction fans. Notwithstanding this physical constraint, quantum teleportation is unique in that it offers *communication as a substitute to transportation*; that is, it does not require moving an object, but rather its reconstruction on a distant platform. Most teleportation protocols to date transfer photon or atom states between distant locations. A teleportation protocol proposed by T. Opatrny (Czech Republic) and G. Kurizki (Israel) in 2001 showed that a molecular state can also be teleported to a distant molecule. This protocol has been generalized by D. Petrosyan (Greece) and G. Kurizki in 2003 to the teleportation of the state of a complex molecule by a set of many measurements whose results contain a large amount of information.

In view of the fascinating possibilities it offers, teleportation may prove to be the major breakthrough of the era of quantum information which is dawning upon us (Chapter 15). But we must tone down our expectations. Decoherence must first be subdued in order to preserve the quantum information shared by the teleportation protocol. Otherwise, teleportation will not progress beyond the simple single-qubit demonstrations we have mastered so far.

14.3 QUANTUM TELEPORTATION AND TRANSMUTATION

The intriguing notion of quantum teleportation conceived by C.Bennett et al. owes its origin to the famed TV series *Startrek*, where a recurring replica is "Beam me up, Scottie." But quantum teleportation must be a disappointment to *Startrek* enthusiasts, since it cannot be carried out without the sharing of classical information (measurement results) at the speed of light between the sender/teleporter and the receiver. By contrast, the *Startrek* version of this protocol takes place instantaneously, irrespective of the distance between the sender and the receiver, thus violating Einstein's prohibition on faster-than-light signaling.

The other difference between the two protocols is that in *Startrek* the tele-ported object vanishes into thin air at the sending node, and materializes from thin air at the receiving node. A realistic protocol would transform the teleported astronaut into a heap of lifeless matter at the sending node, while a similar heap at the receiving node would assume the shape of that astronaut (the teleported person) and be reanimated after the receipt of the radio transmission supplying the measurement results needed to fully characterize the teleported person.

Despite these differences, quantum teleportation has a futuristic flavor to it, as it allows objects to be shipped over great distances only via transmission of the information embodied in the quantum state of the object. It may take a long time to learn how to teleport a molecule, let alone a human. Still, in the distant future, space travel may be partly replaced by large-scale teleportation of objects (or even humans) from earth to distant planets where receivers installed beforehand may exploit the native soil as platforms or substrates for the reconstruction of those objects. The teleportation channel will consist of optical transmission of first entangled photons followed by measurement data from earth to the receiver. Space colonization may then be completely transformed!

Beyond its practical promise, teleportation has intriguing philosophical impli-cations. It separates the essence of the object—its quantum state—from its less essential material substance. Such separation evokes Aristotle's distinction between thr form and matter of an object, or Plato's separation between the idea of an object and its embodiment in the real world of phenomena. Both Aristotle and Plato considered the form or the idea to be eternal, as opposed to their fleeting material counterpart. One may therefore speculate that these philosophers would have been impressed but not shocked to find out that quantum teleportation can transmit the form/idea of an object from one kind of material substrate to another.

If we venture into the mystical world, we may note that teleportation between different material substances is reminiscent of souls' transmutation, an ancient notion shared by many religions—in particular, Hinduism and Buddhism, in which reincarnation plays a central role.

Even without plunging into the mystical world, we may ponder on the deeper significance of quantum teleportation and cryptography. If a multitude of observers—possibly the entirety of mankind—can share both quantum and classical information among them via a giant web of communication, the out-come may be not only an incredible enhancement of information capacity and exchange, but also a collective mind or consciousness that will view the world in ways still unfathomed to we classical individuals (Chapter 15).

Going Places

We are separated and confined
By immense space we cannot traverse.
Were it only that paths we could find
To the far ends of the universe!
Perhaps we shall! Teleportation
May swap us with a distant friend
And bring us to a destination
In a remote, uncharted land.

14.4 APPENDIX: THE QUANTUM TELEPORTATION PROTOCOL

The quantum teleportation protocol involves an input system A, with an arbitrary (unknown) quantum state, an output system C that should receive this arbitrary state, and another auxiliary system B. In Henry's scenario, Henry is the A (sender) system, Eve is the C (receiver) system and Schred is the B (auxiliary) system.

Here we present a quantum teleportation protocol that involves qubits (not people and cats). Accordingly, we will denote the states of all systems by $|0>$s and $|1>$s. Let us describe the initial setting of the protocol—an arbitrary quantum state of system A:

$$|1>_A$$

We now consider the protocol details, stage by stage.

1) The first stage is to create an entangled channel which includes the receiver qubit C and the auxiliary qubit B. Any fully entangled state will suffice, so we select the following:

$$|\Phi>_{BC} = \frac{1}{\sqrt{2}} (|0>_B|0>_C + |1>_B|1>_C)$$

The protocol requires B & C to be fully entangled, and the auxiliary system to be located near A. The latter requirement is needed at the second stage of the protocol.

2) The joint measurement of A & B is performed in their fully entangled basis, also known as the Bell basis, given by:

$$|\Phi_+>=\frac{1}{\sqrt{2}}\left(|00>+|11>\right)$$

$$|\Phi_->=\frac{1}{\sqrt{2}}\left(|00>-|11>\right)$$

$$|\Psi_+>=\frac{1}{\sqrt{2}}\left(|01>+|10>\right)$$

$$|\Psi_->=\frac{1}{\sqrt{2}}\left(|01>-|10>\right)$$

These Bell states span the entire two-qubit space and thus can serve as a measurement basis. To show the rationale behind the entire protocol, we rewrite the three-qubit state of A & B & C in the A & B Bell state space:

$$|\psi>_A\ |\Phi>_{BC} = \frac{1}{\sqrt{2}}(\alpha|0>+\beta|1>)_A(|0>|0>+|1>|1>)_{BC}$$

$$= \frac{1}{2}\ |\Phi_+>_{AB}(\alpha|0>+\beta|1>)_C+$$

$$\frac{1}{2}\ |\Phi_->_{AB}(\alpha|0>-\beta|1>)_C+$$

$$\frac{1}{2}\ |\Psi_+>_{AB}(\beta|0>+\alpha|1>)_C+$$

$$\frac{1}{2}\ |\Psi_->_{AB}(\beta|0>-\alpha|1>)_C$$

In the original formulation, B & C are entangled whereas A is in its arbitrary state. In the new formulation we have spanned the same three qubit states in an A & B entangled-qubits basis. This reformulation has transferred the α,β dependence to C. Importantly, the equality above is simply an identity between two expressions for the same state in different bases, without any physical manipulation.

The joint measurement of A & B in the Bell basis, as any non-trivial measurement, has unknown and random results whose probabilities are dictated by the wavefunction. The measurement collapses the three-qubit wavefunction to a single state. For example, if the measurement results in $|\Psi_+>_{AB}$, then the three-qubit state collapses to: $|\Psi_+>_{AB}(\beta|0>+\alpha|1>)_C$. There is, for example, a 25% chance of measuring each of the Bell states of qubits A & B: $|\Phi_+>_{AB}, |\Phi_->_{AB}, |\Psi_+>_{AB}, |\Psi_->_{AB}$.

The essence of this stage, which is the heart of the teleportation protocol, is the following. Whereas before the measurement C & B were entangled in the channel, after the joint measurement of A & B in the Bell basis, no matter what the result, C becomes disentangled from B, while A & B become entangled instead. In other words, there is entanglement transfer from entangled B–C to entangled A–B. At the same time, the information, encoded in the α, β amplitudes, is transferred from A to C.

3) As can be seen, the state of C after the measurement is *not identical* to the original input state A for all Bell basis results. It is identical if the measurement results in $| \Phi_+ >_{AB}$, but for all others it is rotated. Hence, the last stage of the teleportation protocol is a measurement-dependent rotation:

$$| \Phi_+ >_{AB} \rightarrow I_C(\alpha|0> +\beta|1>)_C = (\alpha|0> + \beta|1>)_C$$
$$| \Phi_- >_{AB} \rightarrow \sigma_z(\alpha|0> -\beta|1>)_C = (\alpha|0> + \beta|1>)_C$$
$$| \Psi_+ >_{AB} \rightarrow \sigma_x(\beta|0> +\alpha|1>)_C = (\alpha|0> + \beta|1>)_C$$
$$| \Psi_- >_{AB} \rightarrow \sigma_z\sigma_x(\beta|0> -\alpha|1>)_C = (\alpha|0> + \beta|1>)_C$$

Here, I_c is the identity operator, whereas σ_x and σ_z are the following qubit-rotation matrices:

$$\sigma_x = \begin{pmatrix} 0 & 1 \\ 1 & 0 \end{pmatrix}, \quad \sigma_z = \begin{pmatrix} 1 & 0 \\ 0 & -1 \end{pmatrix}$$

The "wonder" of teleportation stems from the *non-locality of entanglement*: receiver C may be far away from the sender and auxiliary A & B systems, and yet, as discussed in Chapter 7, when one of two entangled systems is measured, the other (although unmeasured) also immediately collapses. The same is true here. However, for the teleportation protocol to work, system C must be subjected to a rotation that depends on the result of measuring A & B. To this end, the measurement result (telling us which of the four Bell states was measured) must be sent from the measurement location to that of C. Only when this (classical) information arrives at the C location can the appropriate rotation be performed in order to retrieve the (unmeasured) sender state. Without the measurement result information we have a non-selective measurement (Chapter 7), so that the C state must be traced over all possible (unknown) results. This procedure makes the C state fully mixed; that is, it contains no information at all regarding the initial sender state. Thus, for the quantum teleportation protocol to succeed, classical information transfer between the sender and receiver nodes is mandatory. Such classical information requires a time delay of at least the distance between the nodes divided by the speed of light. Prior to the arrival of this information, the receiver (C–) state cannot be meaningfully related to the sender (A–) state.

The Quantum Counter Revolutionaries

254

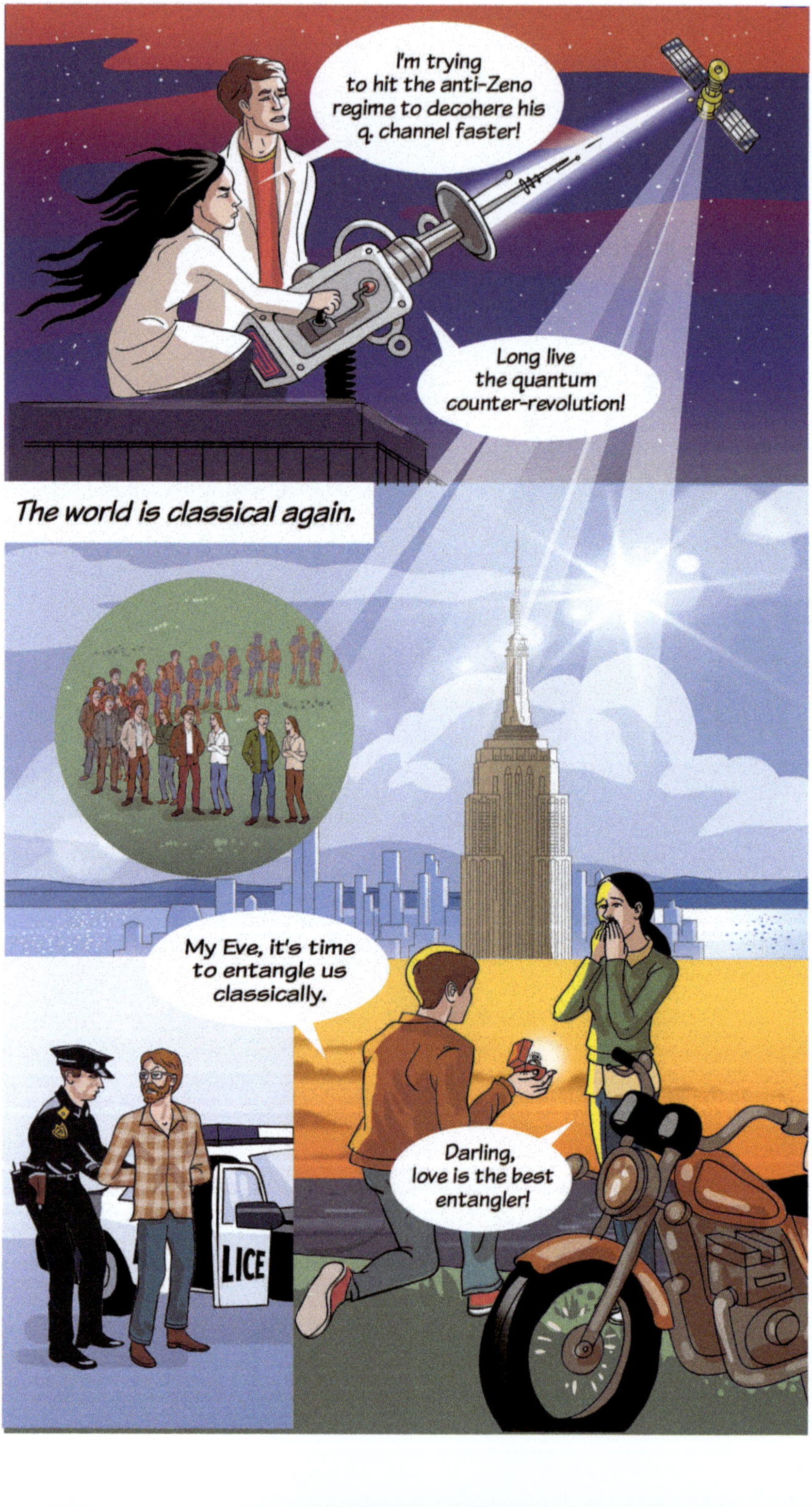

The world is classical again.

CHAPTER 15

The Dawn of Quantum Information

15.1 QUANTUM COMPUTERS: PROMISE AND MENACE

The quantum super-lens has become a formidable technological game-changer in the hands of Henry and Eve: It has opened for them new and hitherto unimaginable possibilities in the realm of quantum technologies based on highly complex systems. Having gauged these possibilities, Henry and Eve have decided to harness all their resources to implement the "holy grail" of quantum technologies: a fully-fledged, large-scale quantum computer.

The glorious day has come: they have finally tested their quantum computer with resounding success! Now they are overcome with joy, but also with a solemn sensation that this is a momentous day for mankind: the dawn of quantum computing has begun.

To realize the magnitude of their achievement, let us recall that present-day "classical" computers process logical bits that can assume the values of EITHER 0 OR 1. By contrast, as we have learnt, quantum mechanics allows a system to be in many states at the same time. The simplest example is a two-level system acting as a *quantum bit* (*qubit*) that can have 0 AND 1 values simultaneously. A computer based on the principles of quantum mechanics can process *both the $|0>$ and $|1>$ states of a qubit at the same time*. Thus, a computation that would take classically 2 seconds (1 second to compute for 0, and another second for 1) would take on a quantum computer only 1 second (one computation for both 0 AND 1 in parallel). This speedup may not look impressive, but consider what happens when the quantum computer has 1,000 qubits. The superposition state of all the qubits has 2^{1000} possible states, which a quantum computer can process *in a single* *step*. Thus, the speedup of a quantum computer is *exponential* in the number of qubits.

Yet this wondrous promise of a quantum computer may only be fulfilled if an enormous hurdle is removed. All the quantum algorithms (computations that are meant to run on a quantum computer) conceived to date require *multi-qubit entanglement* as a means of quantum information processing. If this does not seem to be an insurmountable hurdle, one should recall Henry's adventure in the mine (Chapter 9), where the adverse effects of the environment, caused by interactions of his quantum state with stray objects, resulted in Henry's decoherence. As we have learnt from that adventure, whenever a quantum system decoheres, all its quantum advantages disappear. This is also true of a quantum computer, where one must maintain its multi-qubit coherence or entanglement, otherwise the amazing computational speedup this computer potentially allows will be lost. Unfortunately, the rate at which an entangled multi-qubit state decoheres is *exponentially magnified* with the number of qubits, as fast as (or even faster than) quantum algorithms are executed by the computer. Thus it becomes impossible to maintain 1,000 qubits coherent or entangled throughout the quantum computation, unless measures are taken to counter or correct for decoherence. In principle, this task is tantalizing but not impossible. In the present episode, luckily, the remedy has been found: Henry and Eve have perfected the control of decoherence in multi-qubit systems, on the one hand (Chapter 12), and have mastered the execution of extremely fast entangling operations, on the other, through the use of their quantum super-lens, which is still a thing of the future. The net outcome has been their ability to achieve the breakthrough that the quantum information community has been craving for decades!

Even in the midst of their joyful celebration, Henry and Eve cannot help but think in earnest: what is to be done with their new creation. They would love to apply their quantum computer to much faster and more accurate modelling of complex processes than what is allowed by conventional (classical) computers: the design of chemical reactions tailored to treat diseases that are incurable at present; the prediction of long-term climate trends and their control; simulations of brain processes aimed at understanding consciousness. There is no limit, it seems, to the good that quantum computers can bestow on mankind!

Yet Henry and Eve are both aware of the grave menace to our society should the quantum computer fall into the wrong hands, as such a machine can quickly break all existing codes, since those are safeguarded by the excessively (exponentially) long time required for their deciphering. Now that the tremendous speedup promised by the quantum computer has been achieved, all financial, business or governmental secure data will be free for the taking! The list of nightmarish scenarios is inexhaustible!

While they are contemplating the future of society in the age of quantum computers, their door is smashed with a bang, and unsavory characters, led by their beloved Johnny, rush in. In a matter of seconds, Henry, utterly shocked, is knocked down, and the devices they have toiled so much to build—the quantum super-lens, the quantum computer and the quantum teleporter—are "expropriated for the quantum revolution", by order of Johnny. Henry tries to reason with him: "For quantum sake, Johnny, what's this savagery? Aren't we revolutionary comrades, toiling for the good of humanity?" "Ha! My dear, naïve Henry!" Johnny answers contemptuously. "You have no notion of what the good of humanity means! You and Eve were all along mere instruments of my will, which is infinitely superior to that of all mortals! You've been toiling to execute my plans, which you've mistaken for your own! And no more Johnny, please. I am Commander QT – the Quantum Tesla".

"But what's the point of this farce?" Henry exclaims. "Be courteous, Henry", Commander QT replies gravely. "You're about to witness the new dawn of mankind, which will be converted to an entangled quantum state. Just as Tesla dreamt of endowing mankind with unlimited energy resources, I, Commander QT, will harness the quantum information resources you've developed to impose a collective quantum state on all mankind. No more individuality of the mind, which is the source of all our misfortunes and suffering! The dream of sublime sages, the Buddha, Plotinus, Spinoza, will finally come true! From now on, all humanity will share a collective mind, infinitely wiser and more rational than that of any single person".

"Your fiendish plan has got nothing to do with what those sages dreamt about!" Henry shouts with indignation. "But, even technically, it sounds like a pipedream!"

"Don't underestimate my inventiveness, Henry!", Commander QT bursts in haughty laughter. The social media and cellphones will be my channels for manipulating mankind now that the quantum computer can break all cyber defenses. Mankind is desperately awaiting its salvation, I must hurry!" he concludes hastily, leaving the room with his goons and his precious booty.

Eve rushes to help Henry to his feet. "I've suspected all along that Johnny was hiding his true course from us", she exclaims. "He turned me against you, Darling, probably trying to make us compete more fiercely! Come, we must foil his monstrous plan, there is not a moment to lose!" "But how, my dear?" Henry asks in a baffled tone. "We've got all we need, Darling, to disrupt his quantum information channel; he forgets that decoherence is a two-edged sword, and we may turn it on or off with proper control". "You're brilliant, my Eve!" Henry shouts excitedly. "We'll beat him at his own game!".

They start frantically rushing about the room, searching for the needed gear. Within a few minutes they install a powerful laser hooked to another quantum super-lens that Commander QT did not know about, and turn it against the communication satellite that Commander QT is using for his fateful worldwide transmission of quantum information. They keep increasing the pulse rate of the laser until they reach the anti-Zeno regime, in which they expect the pulses to accelerate the decoherence and decay of Commander QT's collective quantum state (Chapter 10). It is high time for their counteraction, and they watch with horror on their monitor screen how people are becoming collectively entangled worldwide. "Aren't we too late?" Henry asks as the signal generated by the collective quantum state keeps growing. Then, a change occurs: the signal becomes increasingly erratic and noisy. Commander QT's tormented face appears on the screen: He has finally realized that his plan is going amiss, and people whom he has converted to a collective state have reverted to their individual signals. His life's work and aspirations are lying in shambles.

Shortly thereafter, the authorities, alerted by Henry and Eve, apprehend Commander QT and his comrades-in-arms at their headquarters. Watching Johnny's handcuffed low-bent figure being whisked into a police car, Henry's feelings alternate between immense relief and pity. "Oh, what a fall was there", he murmurs to Eve. "Still, let us rejoice: Commander QT's collective tyranny of the mind has failed. Long live the quantum counter-revolution!".

Later that day, Henry asks for Eve's hand in marriage. "Oh, Henry, Darling, why have you taken so long? Let's have the grandest quantum wedding the world has seen, shall we? After all, this is the dawn of the quantum era." "Yes, my Eve, we shall. And then we'll try to save the world with our quantum computer. But it'll have to wait until we come back from our honeymoon." "Where shall we start, Darling? We'll have our hands full, won't we?" "I'm not sure, my Eve. But we'll think of something."

15.2 FROM QUANTUM COMPUTING TO QUANTUM TECHNOLOGY

Even before the advent of quantum teleportation and cryptography there had been speculation concerning the possible application of quantum coherence to computing. The notion of quantum computing, which was first introduced by P. Benioff (Israel) and Y. Manin (USSR) in 1980 and by R. Feynman (USA) in 1982, was based on the idea that a quantum computer obeying the unitarity of QM is capable of acting in parallel on a superposition of two possible logical states, 0 and 1 or spin-down and spin-up states, and is thus equivalent to a massive number of classical computers acting simultaneously in the space of 0 and 1. A superposition state was viewed by the pioneers of quantum computing as a resource for ensuring enormous gains in information capacity and processing speed. D. Deutsch (UK, 1985)—an adept of Everett's many-world interpretation (Chapter 4)—saw quantum computation as correlated (synchronous) processing of information by a classical computer in many worlds, each yielding its own outcome.

The breakthrough in our appreciation for the significance of quantum computing came when P. Shor (USA, 1994) developed his quantum algorithm for the factorization of any large number into prime numbers. The algorithm revealed that the necessary resource for quantum computing is the entanglement of many qubits. This resource can be built up by the entanglement of one pair of qubits at a time, termed a *two-qubit* or *controlled-not* gate operation, followed by the entanglement of each of the two qubits with two other partners, and so on, until all qubits become mutually entangled, operating as a massively entangled register (Figure 15.1). The tremendous expected payoff is the exponential speedup of the factorization algorithm with the number of qubits N, compared to its classical counterparts: the latter can at best shorten the factorization time by some finite power of N. The tremendous practical significance of Shor's algorithm is its ability to greatly speed up the cracking of secret codes encrypted in the factorization of large numbers into prime numbers—a highly time-consuming computation which to date is the standard protection in banking transactions. This possible application of Shor's algorithm has not eluded the attention of governmental and private agencies with vested interests, and has ensured their generous support of research in quantum computing. It has allowed the extensive continuous quest over the past two decades for the realization of quantum computers in a broad variety of physical systems: cold trapped ions or atoms, superconducting circuits or impurities in solids have all been shown to act as qubits whose quantum state is encoded in the electronic or spin state of the system, with the view of employing them as building blocks of a quantum computer.

As shown by A. Barenco et al. (USA, 1995), a combination of one-qubit rotators and two-qubit entanglers allows for universal quantum-computation gates. This means that linearly connected arrays of such elements can be the building blocks of a universal quantum computer, capable of any computation that employs quantum superposed and entangled states for parallel processing, such as Shor's

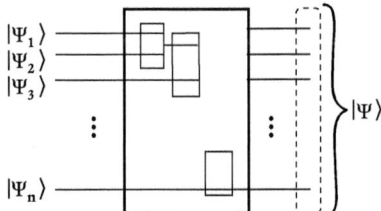

$|\Psi_1\rangle$
$|\Psi_2\rangle$
$|\Psi_3\rangle$

$\vdots$

$|\Psi_n\rangle$

$\Big\}|\Psi\rangle$

Fig. 15.1 Quantum register, creating multi-qubit entanglement (see text).

algorithm. However, the integration of many such building blocks in a quantum computer that would implement Shor's algorithm for numbers so large that their factorization is unfeasible by classical computers has turned out thus far to be an effort that is too ambitious to be managed. The great challenge is the realization of a huge and massively entangled array of qubits by many entangling steps. Each step can be affected by controlled interaction between two qubits, as first shown by I. Cirac and P. Zoller (Austria, 1995) and experimentally demonstrated by D. Wineland (USA, 2002) for qubits embodied by cold trapped ions. And yet the initial enthusiasm regarding the feasibility of scaling up this procedure to large arrays has since been tempered by the recognition that multi-qubit entanglement has a formidable foe: decoherence, which destroys entanglement among qubits faster the greater their number, resulting in computational error.

The remedy for decoherence that has been implemented with some success consists of quantum error correction codes whose essence is as follows. 1) Encode the *logical* qubit by several entangled *physical* qubits; e.g., $|0>\rightarrow |000>, |1>\rightarrow |111>$. 2) Rely on the expectation that error probability is small if decoherence is slower than the computation, then an error will occur on a single *physical* qubit. 3) Perform what is called a *syndrome measurement*, which detects the error but not the value of the *logical* qubit; e.g., if the error is a bit-filp ($0\rightarrow1$, $1\rightarrow0$) then detect if the *physical* qubits are different from one another, and which one has flipped. 4) Perform a correction operator only on the *physical* qubit that has changed.

It has, however, become increasingly clear that other measures—notably, dynamical control of decoherence (Chapter 12)—must be employed to supplement error correction by suppressing or preventing decoherence.

Be that as it may, progress towards a large-scale quantum computer has been painstakingly slow. The latest record of fifty entangled trapped-atom qubits by M. Lukin (USA, 2017) and nearly simultaneously 53-ions quantum simulator by C. Monroe (USA, 2017), is promising (Figure 15.2). It may take a very long time before a 1,000-qubit quantum computer is realized with reliability high enough to outperform conventional (classical) code cracking. As if that were not enough, we are facing the strange situation that Shor's algorithm solely captures the imagination of potential consumers. This prompts the question of whether the

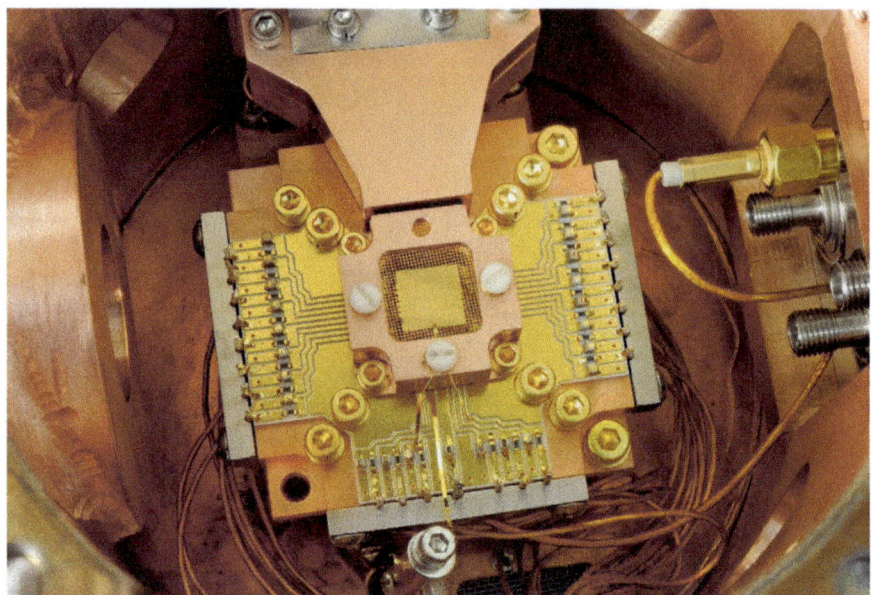

Fig. 15.2 An entangling device for multiple trapped-atom qubits in the setup of Monroe et al. The multiple electrodes allow us to steer addressable trapped-atom pairs such that they become entangled upon meeting and interacting with each other.

colossal effort invested in quantum computing is commensurate with its best-case outcome: a full-scale implementation of Shor's algorithm.

As a "poor man's substitute" for a fully-fledged quantum computer, an alternative has been declared: quantum simulators—machines capable of performing one operation only that mimics some natural quantum process, in the spirit of Feynman's original proposal. Such simulators are the quantum counterparts of classical analog computers that have been obsolete since the 1950s. We may learn very interesting physics from quantum simulators, but their overall technological potential is still unclear.

Forced to be restricted to a small number of reliably entangled qubits for the time being, quantum information scientists have begun to look for other applications where a few entangled qubits may suffice. Certain applications of quantum information processing with small qubit numbers are promising. Among them, quantum sensing, notably, feeble force detection by entangled qubits is particularly appealing: the observability of electric, magnetic or gravitational forces may be optimized by keeping the entangled qubits in quantum states that allow for maximally precise observation. There are hopes that such methods may boost the performance of medical diagnostics (magnetic resonance imaging, MRI, for one), gravitometry (which may facilitate oil prospecting) and more. Here too, decoherence must be suppressed in order to attain optimized observation.

The foregoing survey leads to the conclusion that the major hindrance en route to the development of quantum technologies, ranging from quantum computing through quantum teleportation and cryptography to quantum sensing, is decoherence. If our methods of dynamical control of decoherence live up to their promise, we may some day be able to subdue decoherence to the extent that we can superpose, entangle and teleport highly complex objects such as people and cats. The fantastic tale of Henry, Eve and Schred may then come true.

15.3 LONG LIVE THE QUANTUM REVOLUTION!?

Many politicians, entrepreneurs and scientists alike thrive in anticipation of the second quantum revolution—the advent of quantum technologies that we have surveyed. Let us fantasize that the day will come when all obstacles are removed, and the second quantum revolution emerges victorious. We shall then be able to create massively entangled states of many people—perhaps a quantum web encompassing all mankind. What a glorious day that would be!

But will this revolution be the fulfilment of a happy dream for mankind? We think that such a scenario spells, on the contrary, a horrific nightmare for future generations. The reason is that massive entanglement will deprive people of their individuality, as in a collective Dicke state (Chapter 7). It may be, say, a sum over all possible states, each corresponding to another person being excited (alert) and all the rest unexcited (dormant). As long as such a massively entangled collective state is kept intact, mankind will think and act as one. This is Johnny's (or QT's) goal in his quantum revolution attempt foiled by Henry and Eve. Johnny is sure that he is about to impose on mankind a state of total, permanent happiness.

An yet history abounds in examples of revolutions that pledged happiness but ended up in subjecting people to dismal suffering in an attempt to keep them in line. A quantum-entangled collective state is the ultimate form of order, cohesiveness and therefore lack of freedom. Even the fledgling social network technologies today point in this direction—but the case of a totally entangled state of mankind would be infinitely more extreme!

The saving grace, oddly enough, may come from decoherence, which makes such states extremely fragile (Chapter 8). A flip of the state of one of the superposed individuals, say, from an excited (alert) state to the unexcited state of slumber or indifference would utterly destroy the entire collective (entangled) state. This means that disobedience by any individual may topple this tyranny!

It is important to examine the roots of human yearning for collectivity. Although each ideology charts a different course and a different goal for collectivity, diverse ideologies share the belief that misery comes from individuality and that its suppression is the route to common good. But how far back does this trend go? Although we tend to single out recent or existing ideologies of this kind, very few are aware of their ancient origin: the prime suspect, at least in Western culture, is the doctrine of Plotinus, a respected second-century AD Alexandrian philosopher known as the Father of Neo-Platonic philosophy. Clothed in a perfectly rational disguise of logical thought, this doctrine preached the return of mankind to its pristine, primordial state of collectivity, because this state meant ultimate simplicity and therefore total bliss, which Plotinus equated with total apathy. He wrote: "But who are we? . . . Pure spirits, undifferentiated but unified with the One."

Some 1,500 years later, Spinoza (Holland) put forward a hierarchy of ideas in which collective modes of thought take precedence over individual, fleeting modes. This was, however, an abstract metaphysical system—a far cry from a political or social doctrine. By contrast, the twentieth-century intellectual climate abounded with such doctrines that sought to employ science and technology as tools of *social engineering*. Thus, F. Skinner's influential utopia of social behaviorism (USA, 1948) set the supreme goal of dismissing individuality and enclosing human behavior in what has become known as "Skinner's box". In a sense, Commander QT is Skinner's intellectual descendent. Commander QT's opponents, Henry and Eve, lean towards the fierce individualism of N. Chomsky (USA), Skinner's rival, or that of A. Huxley and G. Orwell (UK), whose dystopias depict utterly regimented and engineered societies.

Let us be vigilant, our readers: science, for all its blessings and wonders, may carry us towards the dangerous shores of collectivity, where our personal freedom is entirely compromised!

Quantum Mankind

A quantum network may some day
Extend entanglement worldwide.
Then, unaware, mankind may
Forfeit its individual side.
Our fears, our doubts, division, strife,
Selfish perception, will be gone.
Confusion will be purged from life

And no one will then feel alone.
The massive thinktank of mankind
May then attain astounding heights!
We're lucky this collective mind
Is still a nightmare, not in sight.

15.4 APPENDIX: A TASTE OF EXPONENTIAL SPEEDUP

Let us present the readers with a flavor of what quantum computers are potentially capable of, by introducing the simple Deutsch–Josza quantum algorithm. It is an elementary example of how quantum computations can solve problems exponentially faster than their classical counterparts.

The problem formulation is as follows. For a number x, represented by a sequence of bits of length n, we are given a function, $f(x)$, that returns either 0 or 1 for each such sequence. We are told that the function is either (i) constant, which means that regardless of the input, it outputs either 0 or 1; or (ii) balanced, meaning that it outputs 0 for one half of the inputs and 1 for the other half, but it is unknown which inputs belong to which half. The goal is to determine whether the function is *constant* or *balanced*.

On a conventional (classical) computer, this algorithm can be extremely slow. The only way to ensure a correct answer in the worst-case scenario is to check more than half of the possible inputs, because only after checking the majority of inputs can we determine whether the function is constant or balanced. For a sequence of n bits, checking more than half the inputs means at least $2^{n-1} + 1$ inputs out of the 2^n possible inputs. This means that the number of evaluations, in the worst-case scenario, is *exponential* in the number of bits. For example, for 1,000 bits we will need $2^{999} + 1$ evaluations, which is an enormously large number.

Can a quantum algorithm do better? Let us first define the problem within a quantum formalism. Given a state of n qubits and another auxiliary qubit, the evaluation of the number, represented by the n qubits, is given by the auxiliary qubit output:

$$f \, | \, x{>}_{(n)} \, | \, 0 > \quad \rightarrow \quad | \, x{>}_{(n)} \, | \, f(x) >$$
$$f \, | \, x{>}_{(n)} \, | \, 1 > \quad \rightarrow \quad | \, x{>}_{(n)} \, | \, 1 - f(x) >$$

Recalling that $f(x)$ can be either 0 or 1, the output qubit holds the value of the function.

The crux of the Deutsch–Jozsa algorithm is to utilize the superposition of *all* possible numbers represented by the qubits. Thus, one prepares a complex input state:

$$| \psi >= \frac{1}{\sqrt{2^{n+1}}} \sum_{x=0}^{2^n} | x > (|0 > -|1 >)$$

We rotate the auxiliary qubit to the $|+ >$ state and superpose all the possible inputs. Applying the function operator on this complex state results in:

$$f | \psi >= \frac{1}{\sqrt{2^{n+1}}} \sum_{x=0}^{2^n} | x > (| f(x) > -|1 - f(x) >)$$

We keep in mind that *f(x)* can be either 0 or 1. If it is 0 the auxiliary state remains the same; if it is 1 the auxiliary qubit changes sign (from $(|0 > - | 1>) \rightarrow (|1> -|0>))$. Hence we can write the state as:

$$f | \psi >= \frac{1}{\sqrt{2^{n+1}}} \sum_{x=0}^{2^n} (-1)^{f(x)} | x > (|0 > -|1 >)$$

The next step is the only complicated one mathematically, so let us first consider the easiest case, known as the Deutch algorithm, for which $n = 1$. In this case we have:

$$f | \psi >= \frac{1}{2} \left((-1)^{f(0)} |0 > + (-1)^{f(1)} |1 > \right) (|0 > -|1 >)$$

We next rotate the first qubit, using a Hadamard operation (Chapter 8):

$$(|0 > \rightarrow \frac{1}{\sqrt{2}} (|0 > + |1 >), |1 > \rightarrow \frac{1}{\sqrt{2}} (|0 > -|1 >)) :$$
$$\rightarrow \frac{1}{2} \left(\left((-1)^{f(0)} + (-1)^{f(1)} \right) |0 > + \left((-1)^{f(0)} - (-1)^{f(1)} \right) |1 > \right) (|0 > -|1 >)$$

We then measure the first qubit: if *f* is constant, then $f(0) = f(1)$, and we obtain $|0>$ with 100% certainty; if *f* is balanced, then $f(0) \neq f(1)$, and we obtain $|1>$ with 100% certainty. We have thus found the solution by a *single* operation of *f* on a single quantum state.

Going back to the original *n*-qubit definition, we perform a Hadamard rotation on each of the *n* qubits. This results in the rather complex state of:

$$\rightarrow \frac{1}{2^n} \sum_{x=0}^{2^n} (-1)^{f(x)} \sum_{y=0}^{2^n-1} (-1)^{x \cdot y} | y > (|0 > -|1 >)$$
$$= \frac{1}{2^n} \sum_{x=0}^{2^n} \left[\sum_{y=0}^{2^n-1} (-1)^{f(x)} (-1)^{x \cdot y} \right] | y > (|0 > -|1 >),$$

where *y* is a representation of each number by a sum of all other numbers, due to the Hadamard rotation, and $x \cdot y$ is the sum of the qubit-by-qubit products. We measure all *n* qubits (ignoring the auxiliary one) and look for the probability of finding the state $y = 0$; i.e., all *n* qubits are equal to zero, and obtain:

$$p\left(y=0\right)=\left|\frac{1}{2^n}\sum_{y=0}^{2^{n-1}}(-1)^{f(x)}\right|^2$$

If f is constant we obtain $p(y=0)=1$, whereas if f is balanced we obtain $p(y=0)=0$. Thus, using this quantum algorithm, we obtain a *deterministic* and correct result, 100% of the time, by a *single* calculation of f.

This example shows that for a carefully designed input state, which includes a special superposition of n qubits, one can construct a set of quantum operators whose action on this state can achieve *exponential speedup* compared to its classical counterpart computation. It is important to note that not any superposition of all possible states in the input will suffice, as the initial state is a superposition of *all possible states*. In addition, we need the ability to implement a set of operators (rotations, two-qubit gates) and a set of multi-qubit (entangling) measurements, as in quantum teleportation (Section 14.4). Not only is the implementation of such tools enormously challenging, mainly because of decoherence, even the quantum algorithm design is still in its infancy, so that quantum software engineering is a challenge of similar magnitude. These challenges are formidable indeed, but not insurmountable, and the potential payoff of exponential computational speedup amply justifies tackling them head-on with diverse technological and theoretical tools.

INDEX

Alpha-particle 24, 26, 52, 228
Anti-Zeno Effect (AZE) 170, 171, 173, 174, 175
Arrow of time 97, 152, 153, 154, 196, 197, 198, 233
Atomism 8, 9, 10, 11, 13, 15, 177
Avogadro's number 9, 11

Bang-Bang 209, 216, 217, 218, 232
Bell measurement 245
Bell state 245, 251, 252
Beta-decay 171, 176
Black holes 34
Blackbody radiation 33
Boltzmann factor 194
Bra-ket formalism 71, 72, 117, 136
Born's superposition princieple 31, 47, 68, 69, 70
Brownian motion 11, 154

Coarse-graining 153, 154
Coherent transfer 143, 145, 157, 159, 166, 167, 168, 169, 170, 181, 182, 183, 184, 190
Complementarity 80, 81, 82, 83, 84, 94, 132, 134
Condensed-matter physics 16
Constructive interference 29, 45, 46, 47, 58, 59, 124, 157, 208, 214, 229, 230
Continuous-variable 84, 85, 87, 88, 101
Copenhagen interpretation 49, 57, 65, 66, 80, 81, 83
Correspondence principle 7, 47, 48
Cosmic-background radiation 148
Cosmological clock 97, 98

Decoherence-free subspace (DFS) 214, 215
Degenerate 126
Density matrix 66, 134, 135, 199, 200, 201, 212, 217, 218
Dephasing 149, 208, 209, 211, 212, 216, 217, 218
Destructive interference 45, 46, 47, 58, 59, 124, 126, 157, 207, 208, 214, 215, 229, 230
Dicke state 265
Discrete variable 84
Duality 13, 46

Dynamical control 210, 214, 215, 231, 232, 263, 265

Eigenfunction, eigenstate, eignvalue 31, 39, 44, 47, 50, 51, 53, 54, 65, 66, 69, 81, 86, 112, 127, 130, 153, 170, 171, 197, 210, 211, 212, 229
Electromagnetism 5, 13, 14, 27
Entanglement Thermalization Hypothesis (ETH) 153
Entropy 50, 56, 66, 97, 98, 115, 131, 150, 152, 153, 154, 194, 195, 197, 198
Epistemology 34, 83
EPR 113, 116

First law of thermodynamics 194

Ghost imaging 233
Golden Rule 171

Hawking radiation 234
Heisenberg uncertainty principle 79
Heisenberg's microscope 81, 82
hidden variables 49, 68
Hydrogen atom 47, 81

Indistinguishable 54, 84, 124, 153, 170, 183, 195, 247

Many-body physics 16
Many-world interpretation 132, 262
Maxwell's laws 6, 12, 13, 24, 27, 94
Molecules 6, 9, 16, 25, 31, 32, 46, 52, 54, 112, 148, 152, 190, 214, 226, 244
Monistic philosophy 84
Multi-qubit 259, 263, 269
Multiverse 70, 132

Non-commutativity 33, 80, 81, 94, 244
Non-locality 116, 252
Normalization 32, 36, 57, 71, 72, 73, 85, 88, 236
Nuclear magnetic resonance (NMR) 51, 52, 210, 211
Nucleus 11, 12, 13, 14, 25, 29, 81, 176, 195, 196, 224, 225, 228

Observable 29, 31, 37, 46, 48, 50, 63, 65, 66, 78, 79, 80, 81, 83, 84, 86, 92, 93, 98, 99, 101, 112, 113, 132, 133, 153, 166, 180, 197, 198, 229, 245
Ontology 34
Orbitals 14, 25, 26, 29

Phase 44, 45, 46, 47, 54, 57, 58, 59, 65, 66, 109, 124, 126, 127, 128, 136, 137, 138, 149, 150, 159, 168, 172, 173, 183, 184, 207, 216, 230
Photoelectric effect 12, 13
Photons 6, 7, 13, 16, 19, 24, 28, 65, 69, 79, 81, 94, 99, 112, 116, 127, 145, 148, 149, 151, 152, 231, 232, 233, 245, 246, 247, 249
Photosynthesis 213
Planck's length 234
Pointer states 152
Position-momentum uncertainty relation 88, 228, 229, 233
Probability amplitude 24, 25, 31, 32, 36, 38, 39, 44, 48, 50, 51, 54, 57, 59, 64, 65, 70, 71, 85, 88, 138, 167, 201
Projection postulate 68, 69, 112
Projective measurement 172

Quanta 6, 12, 13, 14, 15, 19, 24, 27, 31, 81, 99, 147, 149, 150, 171, 176
Quantum coherence 113, 124, 125, 126, 127, 139, 145, 202, 213, 218, 261
Quantum computers 55, 114, 115, 116, 126, 258, 259, 260, 262, 267
Quantum cryptography 213, 245, 246, 247
Quantum Darwinism 67
Quantum error correction 263
Quantum fluctuation 116
Quantum Hologram 116, 117
Quantum Information Processing and Communication (QIPC) 211
Quantum non-demolition measurements (QND) 190, 191, 192, 199, 200, 201, 202
Quantum sensing 264, 265
Quantum teleportation 116, 213, 242, 245, 247, 248, 250, 261, 265, 269
Quantum tunneling 224, 225, 235
Quantum Zeno Effect (QZE) 169, 170, 172, 173, 174, 175, 176, 179, 180, 181, 191, 192,

193, 195, 196, 197, 198, 202, 209, 214, 231, 232, 233

Rabi oscillation 143, 144, 145, 146, 155, 156, 159, 166, 167, 168, 173
Radioactive decay 12, 114, 179, 224, 225, 237

Schroedinger's cat 109, 114
Second law of thermodynamics 150, 153, 194, 196, 213
Secret-key distribution (SKD) 245, 246
Shor's algorithm 262, 263, 264
Solipsism 133
Spin 50, 51, 52, 53, 65, 81, 84, 112, 113, 129, 130, 136, 137, 138, 195, 196, 212, 244, 262
Spin echo 209, 211, 212, 216, 218
Standing waves 14, 46, 47, 236
Statistical mixture 66, 130, 189, 232
Statistical physics 16, 97, 115, 154, 177
Superconducting Josephson junction 226
Superconducting quantum interference devices (SQUID) 55, 226, 228
Superfluidity 7
Superluminal 230, 231, 232
Superradiance 214, 215
System-bath interaction 147, 149, 171, 172, 183

Thermodynamics 10, 12, 13, 16, 56, 97, 98, 150, 154, 193, 195, 196, 202, 213
Time-energy uncertainty relation 94, 95, 174, 175, 179, 192, 197, 199
Trace-out 200
Tunneling electron microscope (TEM) 237
Two-slit interference 28, 54, 82

Virtual quanta 99
Visibility 134, 137, 138, 139, 199

Wavefunction 24, 26, 30, 31, 36, 48, 49, 56, 66, 97, 117, 225, 236
Wavelike 6, 23, 24, 25, 26, 27, 29, 30, 31, 32, 55, 64, 94
Which-path information 124, 125

Zeno's arrow paradox 166, 178